Java Message Service

SECOND EDITION

Java Message Service

Mark Richards, Richard Monson-Haefel, and David A. Chappell

Java Message Service, Second Edition
by Mark Richards, Richard Monson-Haefel, and David A. Chappell

Printed in the United States of America.

Published by O'Reilly Media, Inc., 1005 Gravenstein Highway North, Sebastopol, CA 95472.

O'Reilly books may be purchased for educational, business, or sales promotional use. Online editions are also available for most titles (*http://my.safaribooksonline.com*). For more information, contact our corporate/institutional sales department: 800-998-9938 or *corporate@oreilly.com*.

Editors: Mike Loukides and Julie Steele
Production Editor: Sarah Schneider
Production Services: Appingo, Inc.
Cover Designer: Karen Montgomery
Interior Designer: David Futato
Illustrator: Robert Romano

Printing History:
May 2009: Second Edition.

ISBN: 978-0-596-52204-9

[C]

1242320347

Table of Contents

Foreword

For close to a decade now, I've been a fan of messaging-based systems. They offer a degree of reliability, flexibility, extensibility, and modularity that a traditional RPC or distributed object system simply cannot. Working with them takes a bit of adjustment, because they don't quite behave the same way that an architect or designer expects a traditional n-tier system to behave. This is not to say that they're better or worse; they're just different. Instead of invoking methods on objects directly, where the object can hold conversational state or context, now the message itself has to be self-contained and state-complete.

Which raises an important point.

For any given developer with respect to any given technology, there are four distinct stages.

The first is the Ignorant. We may know the technology exists, or not, but beyond that we remain entirely ignorant about its capabilities. It's a collection of letters, at best, often mentioned in conjunction with other technologies that may or may not matter to what we're doing on a daily basis.

The second is the Explorer. Something piques our curiosity, voluntarily or not. We begin some initial forays into the jungle, perhaps downloading an implementation or reading a few articles. We begin to understand the basic framing of where this thing sits in the broad scheme of things and maybe how it's supposed to work, but our hands-on experience is generally limited to the moral equivalent of "Hello World" and a few other samples.

The third is the Journeyman. After running many of the samples and reading a few articles, we realize that we understand it at a basic level and begin to branch out to writing code with it. We feel reasonably comfortable introducing it into production code and reasonably comfortable debugging the stupid mistakes we'll make with it. We're not experts, by any means, but we can at least get the stuff to compile and run most of the time.

The last, of course, is the Master. After building a few systems and seeing how they react under real-world conditions, we have a deep gestalt with it and can often predict how the tool or technology will react without even running the code. We can see how this thing will interact with other, complementary technologies, and understand how to achieve some truly miraculous results, such as systems that resist network outages or machine failures. When the Java Message Service (JMS) API was first released, back in 1999, before any noncommercial/open-source implementations were available, I distinctly remember looking at it, thinking, "Well, it seems interesting, but it's not something I can use without a real implementation," and setting my printed copy of the specification off to one side for later perusal. My transition to Explorer and Journeyman came a few years later, as I came to understand the power of messaging systems, partly thanks to the few implementations out, partly thanks to my own exploration of other messaging systems (most notably MSMQ and Tibco), but mostly due to the person who wrote this second edition of *Java Message Service*.

I'm still well shy of Master status. Fortunately, both you and I know somebody who is not.

Mark Richards has spent the last several years living the messaging lifestyle, both as an architect and implementor as well as a leader and luminary: the first in his capacity as a consultant, the second in his capacity as a speaker on the No Fluff Just Stuff (NFJS) tour. He has a great "take" on the reasons for and the implications of building message-based systems, and he brings that forth in this nearly complete rewrite of Richard Monson-Haefel and Dave Chappell's first edition. Even if you're in the Ignorant stage of JMS, Mark's careful walkthrough of the basics, through implementation and then the design pros and cons of messaging will bring you to the Journeyman stage fast and leave you with the necessary structure in place to let you reach that Master stage in no time at all.

And that, my friend, is the best anybody can ask of a book.

Happy messaging.

—Ted Neward
Principal Consultant, ThoughtWorks
December 10, 2008, Antwerp, Belgium

Preface

When I was presented with the opportunity to revise *Java Message Service*, I jumped at the chance. The first edition, published by O'Reilly in 2000, was a bestseller and without a doubt the definitive source for JMS and messaging in general at that time. Writing the second edition was an exciting chance to breath new life into an already great book and add new content that was relevant to how we use messaging today. What I failed to fully realize when I took on the project was just how much messaging (or, more precisely, how we use messaging) has changed in the past 10 years. New messaging techniques and technologies have been developed, including message-driven beans (as part of the EJB specification), the Spring messaging framework, Event-Driven Architecture, Service-Oriented Architecture, RESTful JMS interfaces, and the Enterprise Service Bus (ESB), to name a few. The somewhat minor book project that I originally planned quickly turned into a major book project.

My original intent was to preserve as much of the original content as possible in this new edition. However, based on changes to the JMS specification since the first edition was written, as well as the development of new messaging techniques and technologies, the original content quickly shrank. As a result, you will find that roughly 75% of this second edition is new or revised content.

The JMS specification was updated to version 1.1 a couple of years after the printing of the first edition of this book. While not a major change to the JMS specification, the JMS 1.1 specification was nevertheless a significant step toward fixing some of the deficiencies with the original JMS specification. One of the biggest changes in the specification was the joining of the queue and topic API under a unified general API, allowing queues and topics to share the same transactional unit of work. However, the specification change alone was not the only factor that warranted a second edition of the book. As the Java platform has matured, so has the way we think about messaging. From new messaging technologies and frameworks to complex integration and throughput requirements, messaging has changed the way we think about and design systems, particularly over the past 10 years. These factors, combined with the specification changes, are the reasons for the second edition.

With the exception of the `Chat` application found in Chapter 2, all of the sample code has been changed to reflect more up-to-date messaging use cases and to illustrate some additional features of JMS that were not included in the first edition.

I added several new chapters that were not included in the first edition, for obvious reasons. You will find new sections in the first chapter on the JMS API, updated messaging use cases, and a discussion of how messaging has changed how we design systems. You will also find new chapters on message filtering, Java EE and message-driven beans, Spring JMS and message-driven POJOs, and messaging design.

In addition to adding new chapters, I significantly revised the existing chapters. Because I updated the sample code used to illustrate various points throughout the book, I was in turn forced to rewrite much of the corresponding text. This provided me with the opportunity to add additional sections and topics, particularly in Chapter 4, *Point-to-Point Messaging*, and Chapter 5, *Publish-and-Subscribe Messaging*. I also reversed these chapters from the first edition with the belief that it is easier to jump into messaging with the point-to-point messaging model using queues rather than the publish-and-subscribe messaging model using topics and subscribers.

I hope you find the new edition of this book helpful in terms of understanding the Java Message Service and messaging in general.

—Mark Richards

Who Should Read This Book?

This book explains and demonstrates the fundamentals of Java Message Service. It provides a straightforward, no-nonsense explanation of the underlying technology, Java classes and interfaces, programming models, and various implementations of the JMS specification.

Although this book focuses on the fundamentals, it's no "dummy's" book. While the JMS API is easy to learn, the API abstracts fairly complex enterprise technology. Before reading this book, you should be fluent with the Java language and have some practical experience developing business solutions. Experience with messaging systems is not required, but you must have a working knowledge of the Java language.

Organization

The book is organized into 11 chapters and 4 appendixes. Chapter 1 explains messaging systems, messaging use cases, centralized and distributed architectures, and why JMS is important. Chapters 2 through 6 go into detail about developing JMS clients using the two messaging models, point-to-point and publish-and-subscribe, including how to filter messages using message selectors. Chapters 7 and 10 should be considered "advanced topics," covering deployment and administration of messaging systems. Chapter 8 provides an overview of the Java 2, Enterprise Edition (Java EE) with regard to JMS, including coverage of message-driven beans as part of the Enterprise JavaBeans 3.0 specification. Chapter 9 covers the Spring Framework as it relates to messaging. Finally, Chapter 11 provides some insight into many of the design considerations and anti-patterns associated with messaging.

Chapter 1, *Messaging Basics*
: Defines enterprise messaging and common architectures used by messaging vendors. JMS is defined and explained, as are its two programming models, publish-and-subscribe and point-to-point. Many of the use cases and real-world scenarios for messaging are described in this chapter, as are the basics of the JMS API.

Chapter 2, *Developing a Simple Example*
: Walks the reader through the development of a simple publish-and-subscribe JMS client.

Chapter 3, *Anatomy of a JMS Message*
: Provides a detailed examination of the JMS message, the most important part of the JMS API.

Chapter 4, *Point-to-Point Messaging*
: Examines the point-to-point messaging model through the development of a simple borrower and lender JMS application. Also covers some of the finer points of the point-to-point messaging model, including message correlation, dynamic queues, load balancing, and queue browsing.

Chapter 5, *Publish-and-Subscribe Messaging*
: Examines the publish-and-subscribe messaging model through the enhancement of the borrower and lender application developed in Chapter 4. This chapter also covers durable subscribers, nondurable subscribers, dynamic durable subscribers, and temporary topics.

Chapter 6, *Message Filtering*
: Provides a detailed explanation of message filtering using message selectors.

Chapter 7, *Guaranteed Messaging and Transactions*
: Provides an in-depth explanation of advanced topics, including guaranteed messaging, transactions, acknowledgments, message grouping, and failures.

Chapter 8, *Java EE and Message-Driven Beans*
: Provides an overview of the Java 2, Enterprise Edition (Java EE) version 3.0 with regard to JMS and includes coverage of message-driven beans (MDBs).

Chapter 9, *Spring and JMS*
: Provides a detailed explanation of the Spring Framework with regards to JMS, including the Spring JMS Template and message-driven POJOs (MDPs).

Chapter 10, *Deployment Considerations*
: Provides an in-depth examination of features and issues that should be considered when choosing vendors and deploying JMS applications.

Chapter 11, *Messaging Design Considerations*
: Provides insight into and explanation of several design considerations, including the use of internal versus external destinations, request/reply processing, and a discussion of some of the more common messaging anti-patterns.

Appendix A, *The Java Message Service API*
: Provides a quick reference to the classes and interfaces defined in the JMS package.

Appendix B, *Message Headers*
: Provides detailed information about message headers.

Appendix C, *Message Properties*
: Provides detailed information about message properties.

Appendix D, *Installing and Configuring ActiveMQ*
: Provides detailed information about installing and configuring ActiveMQ to run the examples in this book.

Software and Versions

This book covers Java Message Service version 1.1. It uses Java language features from the Java 6 platform. Because the focus of this book is to develop vendor-independent JMS clients and applications, I have stayed away from proprietary extensions and vendor-dependent idioms. Any JMS-compliant provider can be used with this book; you should be familiar with that provider's specific installation, deployment, and run-time management procedures to work with the examples. To find out the details of installing and running JMS clients for a specific JMS provider, consult your JMS vendor's documentation; these details aren't covered by the JMS specification. We have provided the details for running the examples with ActiveMQ, a popular open source JMS provider, in Appendix D.

The source code examples and explanation in Chapter 8 refer to the Enterprise Java-Beans 3.0 (EJB 3) specification. The source code examples and explanation in Chapter 9 refer to version 2.5 of the Spring Framework

The examples developed in this book are available through the book's catalog page at *http://oreilly.com/catalog/9780596522049/examples*. These examples are organized by chapter. Special source code modified for specific vendors is also provided. These vendor-specific examples include a *readme.txt* file that points to documentation for downloading and installing the JMS provider, as well as specific instructions for setting up the provider for each example.

Conventions Used in This Book

The following typographical conventions are used in this book:

Italic
: Used for filenames, pathnames, hostnames, domain names, URLs, email addresses, and new terms when they are defined.

`Constant width`
: Used for code examples and fragments, class, variable, and method names, Java keywords used within the text, SQL commands, table names, column names, and XML elements and tags.

`Constant width bold`
: Used for emphasis in some code examples.

`Constant width italic`
: Used to indicate text that is replaceable.

This icon signifies a tip, suggestion, or general note.

The term *JMS provider* is used to refer to a vendor that implements the JMS API to provide connectivity to its enterprise messaging service. The term *JMS client* refers to Java components or applications that use the JMS API and a JMS provider to send and receive messages. *JMS application* refers to any combination of JMS clients that work together to provide a software solution.

Using Code Examples

This book is here to help you get your job done. In general, you may use the code in this book in your programs and documentation. You do not need to contact us for permission unless you're reproducing a significant portion of the code. For example, writing a program that uses several chunks of code from this book does not require permission. Selling or distributing a CD-ROM of examples from O'Reilly books does require permission. Answering a question by citing this book and quoting example code does not require permission. Incorporating a significant amount of example code from this book into your product's documentation does require permission.

We appreciate, but do not require, attribution. An attribution usually includes the title, author, publisher, and ISBN. For example, "*Java Message Service*, Second Edition, by Mark Richards, Richard Monson-Haefel, and David A. Chappell. Copyright 2009 Mark Richards, 978-0-596-52204-9."

If you feel your use of code examples falls outside fair use or the permission given above, feel free to contact us at *permissions@oreilly.com*.

Safari® Books Online

When you see a Safari® Books Online icon on the cover of your favorite technology book, that means the book is available online through the O'Reilly Network Safari Bookshelf.

Safari offers a solution that's better than e-books. It's a virtual library that lets you easily search thousands of top tech books, cut and paste code samples, download chapters, and find quick answers when you need the most accurate, current information. Try it for free at *http://my.safaribooksonline.com/*.

How to Contact Us

We have tested and verified the information in this book to the best of our ability, but you may find that features have changed (or even that we have made mistakes!). Please let us know about any errors you find, as well as your suggestions for future editions, by writing to:

O'Reilly Media, Inc.
1005 Gravenstein Highway North
Sebastopol, CA 95472
800-998-9938 (in the United States or Canada)
707-829-0515 (international or local)
707-829-0104 (fax)

We have a web page for this book, where we list errata, examples, or any additional information. You can access this page at:

http://www.oreilly.com/catalog/9780596522049/

To comment or ask technical questions about this book, send email to:

bookquestions@oreilly.com

For more information about our books, conferences, software, Resource Centers, and the O'Reilly Network, see our website at:

http://www.oreilly.com

Acknowledgments

These acknowledgments are from Mark Richards and refer to the second edition of this book.

No one ever writes a book alone; rather, it is the hard work of many people working together that produces the final result. There are many people I would like to acknowledge and thank for the hard work and support they provided during the project.

First, I would like to recognize and thank my editor, Julie Steele, for putting up with me during the project and doing such a fantastic job editing, coordinating, and everything else involved with getting this book to print. I would also like to thank Richard Monson-Haefel for doing such a great job writing the first edition of this book (along with David Chappell), and for providing me with the opportunity to write the second edition.

To my good friend and colleague, Ted Neward, I want to thank you for writing the Foreword to this book during your very busy travel schedule and for providing me with insight and guidance throughout the project. Your suggestions and guidance helped bring this new edition together. I also want to thank my friends, Neal Ford, Scott Davis, Venkat Subramaniam, Brian Sletten, David Bock, Nate Shutta, Stuart Halloway, Jeff Brown, Ken Sipe, and all the other No Fluff Just Stuff (NFJS) gang, for your continued support, lively discussions, and camaraderie both during and outside the NFJS conferences. You guys are the greatest.

I also want to thank the many expert technical reviewers who helped ensure that the material was technically accurate, including Ben Messer, a super software engineer and technical expert; Tim Berglund, principle software developer and owner of the August Technology Group, LLC; Christian Kenyeres, principle technical architect at Collaborative Consulting, LLC; and last (but certainly not least), Ken Yu and Igor Polevoy. I know it wasn't easy editing and reviewing the manuscript during the holiday season (bad timing on my part, I'm afraid), but your real-world experience, advice, comments, suggestions, and technical editing helped make this a great book.

To the folks at the Macallan Distillery in Scotland, thank you for making the best single malt Scotch in the world. It helped ease the pain during those long nights of writing, especially during the winter months.

Finally, I would like to acknowledge and thank my lovely wife, Rebecca, for her continued support throughout this book project. You mean the world to me, Rebecca, and always will.

Acknowledgments from the First Edition

These acknowledgments are carried over from the first edition of this book and are from the original authors, Richard Monson-Haefel and David A. Chappell.

While there are only two names on the cover of this book, the credit for its development and delivery is shared by many individuals. Michael Loukides, our editor, was pivotal to the success of this book. Without his experience, craft, and guidance, this book would not have been possible.

Many expert technical reviewers helped ensure that the material was technically accurate and true to the spirit of the Java Message Service. Of special note are Joseph Fialli, Anne Thomas Manes, and Chris Kasso of Sun Microsystems; Andrew Neumann and

Giovanni Boschi of Progress; Thomas Haas of Softwired; Mikhail Rizkin of International Systems Group; and Jim Alateras of ExoLab. The contributions of these technical experts are critical to the technical and conceptual accuracy of this book. They brought a combination of industry and real-world experience to bear and helped to make this the best book on JMS published today.

Thanks also to Mark Hapner of Sun Microsystems, the primary architect of Java 2, Enterprise Edition, who answered several of our most complex questions. Thanks to all the participants in the JMS-INTEREST mailing list hosted by Sun Microsystems for their interesting and informative postings.

Special appreciation goes to George St. Maurice of the SonicMQ tech writing team for his participation in organizing the examples for the O'Reilly website.

Finally, the most sincere gratitude must be extended to our families. Richard Monson-Haefel thanks his wife, Hollie, for supporting and assisting him through yet another book. Her love makes everything possible. David Chappell thanks his wife, Wendy, and their children, Dave, Amy, and Chris, for putting up with him during this endeavor.

David Chappell would also like to thank some of the members of the Progress SonicMQ team—Bill Wood, Andy Neumann, Giovanni Boschi, Christine Semeniuk, David Grigglestone, Bill Cullen, Perry Yin, Kathy Guo, Mitchell Horowitz, Greg O'Connor, Mike Theroux, Ron Rudis, Charlie Nuzzolo, Jeanne Abmayr, Oriana Merlo, and George St. Maurice—for helping to ensure that the appropriate topics were addressed, and addressed accurately. And special thanks to George Chappell for helping him with "split infinitives."

CHAPTER 1

Messaging Basics

Over the years, systems have grown significantly in terms of complexity and sophistication. The need to have systems with better reliability, increased scalability, and more flexibility than in the past has given rise to more complex and sophisticated architectures. In response to this increased demand for better and faster systems, architects, designers, and developers have been leveraging messaging as a way of solving these complex problems.

Messaging has come a long way since the first edition of this book was published in 2000, particularly with respect to the Java platform. Although the Java Message Service (JMS) API hasn't changed significantly since its introduction in 1999, the way messaging is used has. Messaging is widely used to solve reliability and scalability issues, but it is also used to solve a host of other problems encountered with many business and nonbusiness applications.

Heterogeneous integration is one primary area where messaging plays a key role. Whether it be through mergers, acquisitions, business requirements, or simply a change in technology direction, more and more companies are faced with the problem of integrating heterogeneous systems and applications within and across the enterprise. It is not unusual to encounter a myriad of technologies and platforms within a single company or division consisting of Java EE, Microsoft .NET, Tuxedo, and yes, even CICS on the mainframe.

Messaging also offers the ability to process requests asynchronously, providing architects and developers with solutions for reducing or eliminating system bottlenecks, and increasing end user productivity and overall system scalability. Since messaging provides a high degree of decoupling between components, systems that utilize messaging also provide a high degree of architectural flexibility and agility.

Application-to-application messaging systems, when used in business systems, are generically referred to as enterprise messaging systems, or Message-Oriented Middleware (MOM). Enterprise messaging systems allow two or more applications to exchange information in the form of messages. A message, in this case, is a self-contained package of business data and network routing headers. The business data contained in a message can be anything—depending on the business scenario—and usually contains information about some business transaction. In enterprise messaging systems, messages inform an application of some event or occurrence in another system.

Using Message-Oriented Middleware, messages are transmitted from one application to another across a network. Enterprise middleware products ensure that messages are properly distributed among applications. In addition, these products usually provide fault tolerance, load balancing, scalability, and transactional support for enterprises that need to reliably exchange large quantities of messages.

Enterprise messaging vendors use different message formats and network protocols for exchanging messages, but the basic semantics are the same. An API is used to create a message, load the application data (message payload), assign routing information, and send the message. The same API is used to receive messages produced by other applications.

In all modern enterprise messaging systems, applications exchange messages through virtual channels called *destinations*. When a message is sent, it's addressed to a destination (i.e., queue or topic), not a specific application. Any application that subscribes or registers an interest in that destination may receive the message. In this way, the applications that receive messages and those that send messages are decoupled. Senders and receivers are not bound to each other in any way and may send and receive messages as they see fit.

All enterprise messaging vendors provide application developers with an API for sending and receiving messages. While a messaging vendor implements its own networking protocols, routing, and administration facilities, the basic semantics of the developer API provided by different vendors are the same. This similarity in APIs makes the Java Message Service possible.

JMS is a vendor-agnostic Java API that can be used with many different enterprise messaging vendors. JMS is analogous to JDBC in that application developers reuse the same API to access many different systems. If a vendor provides a compliant service provider for JMS, the JMS API can be used to send and receive messages to that vendor. For example, you can use the same JMS API to send messages using SonicMQ that you would using IBM's WebSphere MQ. It is the purpose of this book to explain how enterprise messaging systems work and, in particular, how JMS is used with these systems. The second edition of this book focuses on JMS 1.1, the latest version of the specification, which was introduced in March 2002.

The rest of this chapter explores enterprise messaging and JMS in more detail, so that you have a solid foundation with which to learn about the JMS API and messaging concepts in later chapters. The only assumption we make in this book is that you are already familiar with the Java programming language.

The Advantages of Messaging

As stated at the beginning of this chapter, messaging solves many architectural challenges such as heterogeneous integration, scalability, system bottlenecks, concurrent processing, and overall architecture flexibility and agility. This section describes the more common advantages and uses for JMS and messaging in general.

Heterogeneous Integration

The communication and integration of heterogeneous platforms is perhaps the most classic use case for messaging. Using messaging you can invoke services from applications and systems that are implemented in completely different platforms. Many open source and commercial messaging systems provide seamless connectivity between Java and other languages and platforms by leveraging an integrated message bridge that converts a message sent using JMS to a common internal message format. Examples of these messaging systems include ActiveMQ (open source) and IBM WebSphere MQ (commercial). Both of these messaging systems support JMS, but they also expose a native API for use by non-Java messaging clients (such as C and C++). The key point here is that, depending on the vendor, it is possible to use JMS to communicate to non-Java or non-JMS messaging clients.

Historically, there have been many ways of tackling the issue of heterogeneous systems integration. Some earlier solutions involved the transfer of information through FTP or some other file transfer means, including the classic "sneakernet" method of carrying a diskette or tape from one machine to another. Using a database to share information between two heterogeneous systems or applications is another common approach that is still widely used today. Remote Procedure Call, or RPC, is yet another way of sharing both data and functionality between disparate systems. While each of these solutions have their advantages and disadvantages, only messaging provides a truly decoupled solution allowing both data and functionality to be shared across applications or subsystems. More recently, Web Services has emerged as another possible solution for integrating heterogeneous systems. However, lack of reliability for web services make messaging a better integration choice.

Reduce System Bottlenecks

System and application bottlenecks occur whenever you have a process that cannot keep up with the rate of requests made to that process. A classic example of a system bottleneck is a poorly tuned database where applications and processes wait until database connections are available or database locks free up. At some point the system backs up, response time gets worse, and eventually requests start timing out.

A good analogy of a system bottleneck is pouring water into a funnel. The funnel becomes a bottleneck because it can only allow a certain amount of water to pass through. As the amount of water entering the funnel increases, the funnel eventually overflows because water cannot exit the funnel fast enough to handle the increased flow. IT systems work in much the same way: some components can only handle a limited number of requests and can quickly become bottlenecks.

Going back to our example, if a single funnel can "process" one liter of water per minute, but three liters of water are entering the funnel, the funnel will eventually back up and overflow. However, by adding two more funnels to the process, we can now theoretically "process" three liters of water per minute, thereby keeping up with the demand. Similarly, within IT systems messaging can be used to reduce or even eliminate system bottlenecks. Rather than have requests backing up one behind the other while a synchronous component is processing them, the requests are sent to a messaging system that distributes the requests to multiple message listener components. In this manner the bottlenecks experienced with a single synchronous point-to-point connection are reduced or in some cases completely eliminated.

Increase Scalability

Much in the same way that messaging reduces system bottlenecks, it can also be used to increase the overall scalability and throughput of a system, effectively reducing the response time as well. Scalability in messaging systems is achieved by introducing multiple message receivers that can process different messages concurrently. As messages stack up waiting to be processed, the number of messages in the queue, or what is otherwise known as the *queue depth*, starts to increase. As the queue depth increases, system response time increases and throughput decreases. One way to increase the scalability of a system is to add multiple concurrent message listeners to the queue (similar to what we did in the funnel example previously) to process more requests concurrently.

Another way to increase the overall scalability of a system is to make as much of the system asynchronous as possible. Decoupling components in this manner allows for systems to grow horizontally, with hardware resources being the main limiting factor. However, while this may seem like a silver bullet, the middleware can only be horizontally scaled within practical limits of another major system bottleneck—the database. You can have hundreds or even thousands of message listeners on a single queue providing the ability to process many messages at the same time, but the database may only be able to process a limited number of concurrent requests. Although there are complicated techniques for addressing the database bottleneck issue, the reality is that there will always be practical limits to how far you can scale the middleware layer.

Increase End User Productivity

The use of asynchronous messaging can also increase end user productivity. Consider the case where an end user makes a request to the system from a web-based or desktop user interface that takes several minutes to run. During that time the end user is waiting for the results, unable to do any additional work. By using asynchronous messaging, the end user can make a request to the system and get an immediate response back indicating that the request was accepted. The end user now continues to do other work on the system while the long running request is executing. Once the request has completed, the end user is notified that the request has been processed and the results are delivered to the end user. By using messaging, the end user is able to get more work done with less wait time, making that end user more productive.

Many front-office trading systems use this sort of messaging strategy between the trading application and the backend systems. This type of messaging-based architecture allows the trader to perform other work without having to wait for a response from the system. The trade-off for this increased flexibility and productivity, however, is added complexity. A good architect will always look for opportunities to make various aspects of a system asynchronous, whether it be between a user interface and a system or between internal components within the system.

Architecture Flexibility and Agility

The use of messaging as part of an overall enterprise architecture solution allows for greater architectural flexibility and agility. These qualities are achieved through the use of abstraction and decoupling. With messaging, subsystems, components, and even services can be abstracted to the point where they can be replaced with little or no knowledge by the client components.

Architectural agility is the ability to respond quickly to a constantly changing environment. By using messaging to abstract and decouple components, one can quickly respond to changes in software, hardware, and even business changes. The ability to swap out one system for another, change a technology platform, or even change a vendor solution without affecting the client applications can be achieved through abstraction using messaging. Through messaging, the message producer, or client component, does not know which programming language or platform the receiving component is written in, where the component or service is located, what the component or service implementation name is, or even the protocol used to access that component or service. It is by means of these levels of abstraction that we are able to more easily replace components and subsystems, thereby increasing architectural agility.

Enterprise Messaging

Enterprise messaging is not a new concept. Messaging products such as IBM WebSphere MQ, SonicMQ, Microsoft Message Queuing (MSMQ), and TIBCO Rendezvous have been in existence for many years. Recently, several open source messaging products such as ActiveMQ have entered the market and are being used in enterprise production environments. Also, the introduction of Service-Oriented Architecture (SOA) has given rise to a new type of messaging product known as an Enterprise Service Bus (ESB). Although most ESBs allow for HTTP-based communications, messaging-based communication continues to remain the standard in most production enterprise systems.

A key concept of enterprise messaging is that messages are delivered *asynchronously* from one system to others over a network. To deliver a message asynchronously means the sender is not required to wait for the message to be received or handled by the recipient; it is free to send the message and continue processing. Asynchronous messages are treated as autonomous units—each message is self-contained and carries all of the data and state needed by the business logic that processes it.

In asynchronous messaging, applications use a simple API to construct a message, then hand it off to the Message-Oriented Middleware for delivery to one or more intended recipients (see Figure 1-1). A message is a package of business data that is sent from one application to another over the network. The message should be self-describing in that it should contain all the necessary context to allow the recipients to carry out their work independently.

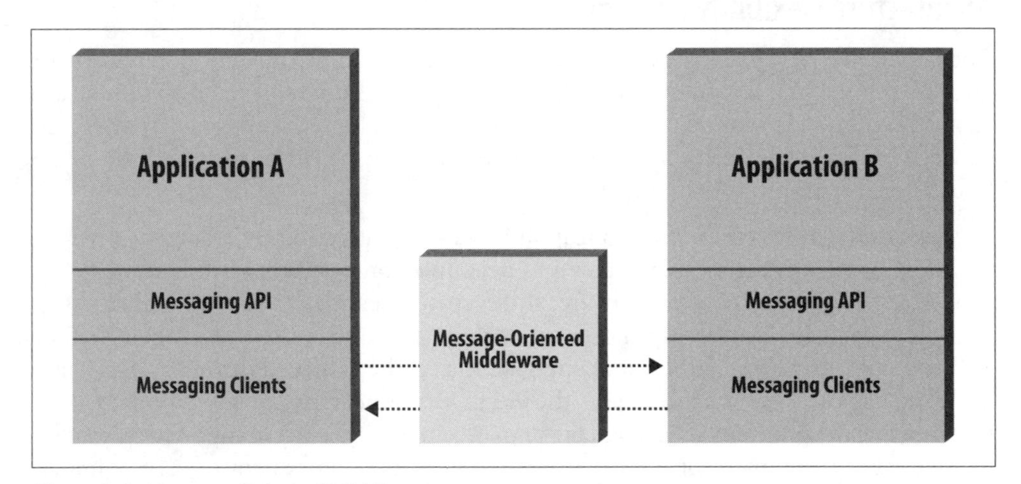

Figure 1-1. Message-Oriented Middleware

Message-Oriented Middleware architectures of today vary in their implementation. The spectrum ranges from a centralized architecture that depends on a message server to perform routing, to a decentralized architecture that distributes the "server" processing out to the client machines. A varied array of protocols including TCP/IP, HTTP, SSL, and IP multicast are employed at the network transport layer. Some messaging products use a hybrid of both approaches, depending on the usage model.

It is important to explain what we mean by the term *client*. Messaging systems are composed of *messaging clients* and some kind of messaging middleware server. The clients send messages to the messaging server, which then distributes those messages to other clients. The client is a business application or component that is using the messaging API (in our case, JMS).

Centralized Architectures

Enterprise messaging systems that use a centralized architecture rely on a *message server*. A message server, also called a message router or broker, is responsible for delivering messages from one messaging client to other messaging clients. The message server decouples a sending client from other receiving clients. Clients see only the messaging server, not other clients, which allows clients to be added and removed without affecting the system as a whole.

Typically, a centralized architecture uses a hub-and-spoke topology. In a simple case, there is a centralized message server and all clients connect to it. As shown in Figure 1-2, the hub-and-spoke architecture lends itself to a minimal amount of network connections while still allowing any part of the system to communicate with any other part of the system.

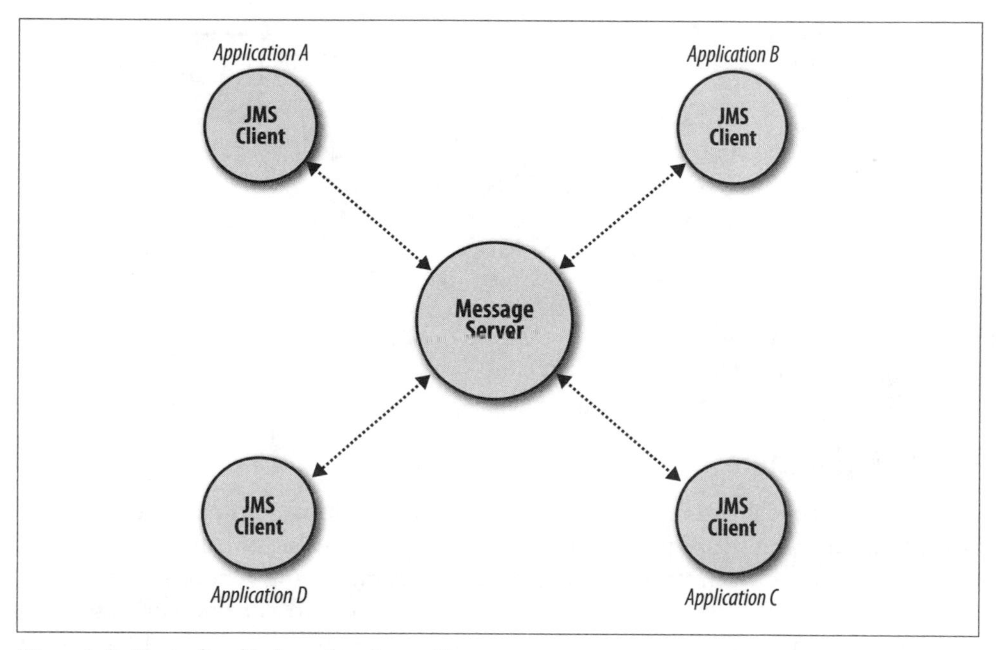

Figure 1-2. Centralized hub-and-spoke architecture

In practice, the centralized message server may be a cluster of distributed servers operating as a logical unit.

Decentralized Architectures

All decentralized architectures currently use IP multicast at the network level. A messaging system based on multicasting has no centralized server. Some of the server functionality (persistence, transactions, security) is embedded as a local part of the client, while message routing is delegated to the network layer by using the IP multicast protocol.

IP multicast allows applications to join one or more IP multicast groups; each group uses an IP network address that will redistribute any messages it receives to all members in its group. In this way, applications can send messages to an IP multicast address and expect the network layer to redistribute the messages appropriately (see Figure 1-3). Unlike a centralized architecture, a distributed architecture doesn't require a server for the purposes of routing messages—the network handles routing automatically. However, other server-like functionality is still required to be included with each client, such as message persistence and message delivery semantics like once-and-only-once delivery.

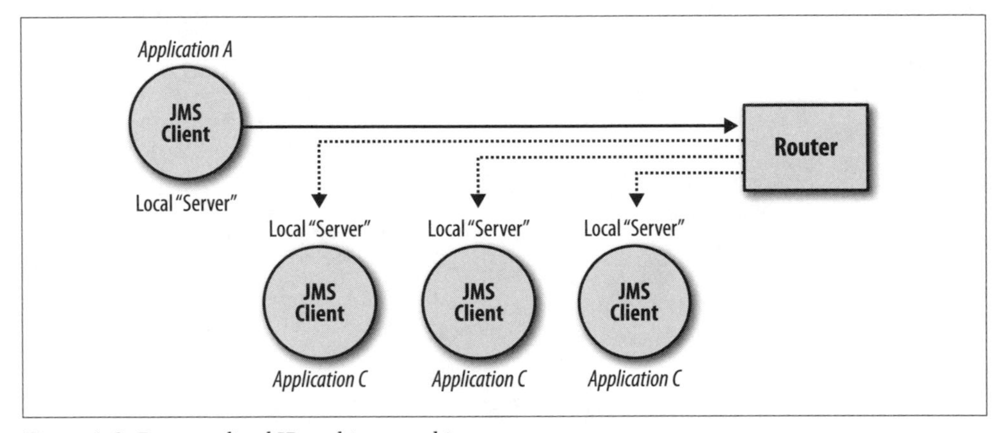

Figure 1-3. Decentralized IP multicast architecture

Hybrid Architectures

A decentralized architecture usually implies that an IP multicast protocol is being used. A centralized architecture usually implies that the TCP/IP protocol is the basis for communication between the various components. A messaging vendor's architecture may also combine the two approaches. Clients may connect to a daemon process using TCP/IP, which in turn communicates with other daemon processes using IP multicast groups.

Centralized Architecture As a Model

Both ends of the decentralized and centralized architecture spectrum have their place in enterprise messaging. The advantages and disadvantages of distributed versus centralized architectures are discussed in more detail in Chapter 10. In the meantime, we need a common model for discussing other aspects of enterprise messaging. To simplify discussions, this book uses a centralized architecture as a logical view of enterprise messaging. This is for convenience only and is not an endorsement of centralized over decentralized architectures. The term *message server* is frequently used in this book to refer to the underlying architecture that is responsible for routing and distributing messages. In centralized architectures, the message server is a middleware server or cluster of servers. In decentralized architectures, the server refers to the local server-like facilities of the client.

Messaging Models

JMS supports two types of messaging models: point-to-point and publish-and-subscribe. These messaging models are sometimes referred to as *messaging domains*. Point-to-point messaging and publish-and-subscribe messaging are frequently shortened to p2p and pub/sub, respectively. This book uses both the long and short forms throughout.

In the simplest sense, publish-and-subscribe is intended for a one-to-many broadcast of messages, while point-to-point is intended for one-to-one delivery of messages (see Figure 1-4).

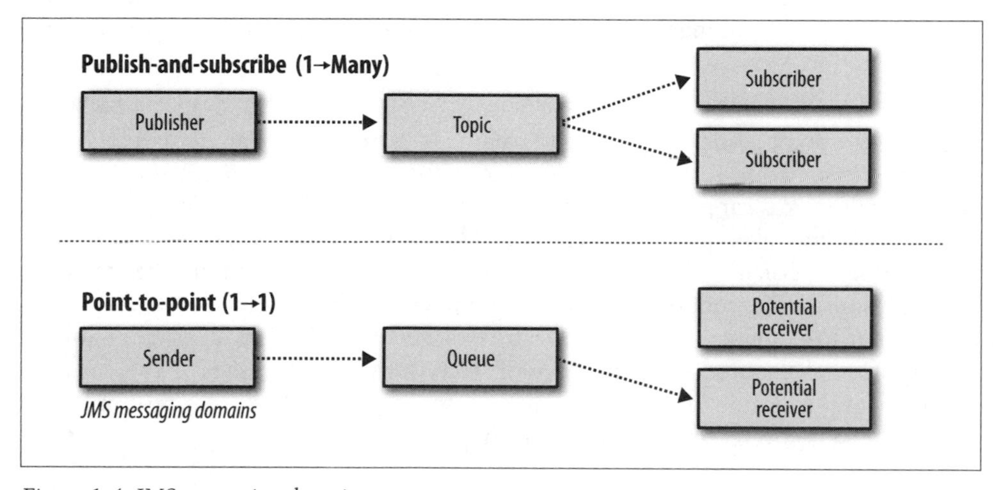

Figure 1-4. JMS messaging domains

From a JMS perspective, messaging clients are called *JMS clients*, and the messaging system is called the *JMS provider*. A *JMS application* is a business system composed of many JMS clients and, generally, one JMS provider.

In addition, a JMS client that produces a message is called a *message producer*, while a JMS client that receives a message is called a *message consumer*. A JMS client can be both a message producer and a message consumer. When we use the term *consumer* or *producer*, we mean a JMS client that consumes messages or produces messages, respectively. We use this terminology throughout the book.

Point-to-Point

The point-to-point messaging model allows JMS clients to send and receive messages both synchronously and asynchronously via virtual channels known as *queues*. In the point-to-point model, message producers are called *senders* and message consumers are called *receivers*. The point-to-point messaging model has traditionally been a pull-based or polling-based model, where messages are requested from the queue instead of being pushed to the client automatically. One of the distinguishing characteristics of point-to-point messaging is that messages sent to a queue are received by one and only one receiver, even though there may be many receivers listening on a queue for the same message.

Point-to-point messaging supports asynchronous "fire and forget" messaging as well as synchronous request/reply messaging. Point-to-point messaging tends to be more coupled than the publish-and-subscribe model in that the sender generally knows how the message is going to be used and who is going to receive it. For example, a sender may send a stock trade order to a queue and wait for a response containing the trade confirmation number. In this case, the message sender knows that the message receiver is going to process the trade order. Another example would be an asynchronous request to generate a long-running report. The sender makes the request for the report, and when the report is ready, a notification message is sent to the sender. In this case, the sender knows the message receiver is going to pick up the message and create the report.

The point-to-point model supports load balancing, which allows multiple receivers to listen on the same queue, therefore distributing the load. As shown in Figure 1-4, the JMS provider takes care of managing the queue, ensuring that each message is consumed once and only once by the next available receiver in the group. The JMS specification does not dictate the rules for distributing messages among multiple receivers, although some JMS vendors have chosen to implement this as a load balancing capability. Point-to-point also offers other features, such as a queue browser that allows a client to view the contents of a queue prior to consuming its messages—this browser concept is not available in the publish-and-subscribe model. The point-to-point messaging model is covered in more detail in Chapter 4.

Publish-and-Subscribe

In the publish-and-subscribe model, messages are published to a virtual channel called a *topic*. Message producers are called *publishers*, whereas message consumers are called *subscribers*. Unlike the point-to-point model, messages published to a topic using the publish-and-subscribe model can be received by multiple subscribers. This technique is sometimes referred to as *broadcasting* a message. Every subscriber receives a copy of each message. The publish-and-subscribe messaging model is by and large a push-based model, where messages are automatically broadcast to consumers without them having to request or poll the topic for new messages.

The pub/sub model tends to be more decoupled than the p2p model in that the message publisher is generally unaware of how many subscribers there are or what those subscribers do with the message. For example, suppose a message is published to a topic every time an exception occurs in a Java application. The responsibility of the publisher is to simply broadcast that an exception occurred. The publisher does not know or generally care how that message will be used. For example, there may be subscribers that send an email to the development or support staff based on the exception, subscribers that accumulate counts of the various types of exceptions for reporting purposes, or even subscribers that use the information to page an on-call support person based on the exception type.

There are many different types of subscribers within the pub/sub messaging model. Nondurable subscribers are temporary subscriptions that receive messages only when they are actively listening on the topic. Durable subscribers, on the other hand, will receive a copy of every message published, even if they are "offline" when the message is published. There is also the notion of dynamic durable subscribers and administered durable subscribers. The publish-and-subscribe messaging model is discussed in greater detail in Chapters 2 and 5.

JMS API

JMS is an API for enterprise messaging created by Sun Microsystems through JSR-914. JMS is not a messaging system itself; it's an abstraction of the interfaces and classes needed by messaging clients when communicating with messaging systems. In the same way that JDBC abstracts access to relational databases and JNDI abstracts access to naming and directory services, JMS abstracts access to messaging providers. Using JMS, an application's messaging clients are portable across messaging server products.

The creation of JMS was an industry effort. Sun Microsystems took the lead on the spec and worked very closely with the messaging vendors throughout the process. The initial objective was to provide a Java API for connectivity to enterprise messaging systems. However, this changed to the wider objective of supporting messaging as a first-class Java-distributed computing paradigm equal with RPC-based systems such as CORBA and Enterprise JavaBeans. Mark Hapner, the JMS spec lead at Sun Microsystems, explained:

> There were a number of MOM vendors that participated in the creation of JMS. It was an industry effort rather than a Sun effort. Sun was the spec lead and did shepherd the work but it would not have been successful without the direct involvement of the messaging vendors. Although our original objective was to provide a Java API for connectivity to MOM systems, this changed over the course of the work to a broader objective of supporting messaging as a first class Java distributed computing paradigm on equal footing with RPC.

The result is a best-of-breed, robust specification that includes a rich set of message delivery semantics, combined with a simple yet flexible API for incorporating messaging into applications. The intent was that in addition to new vendors, existing messaging vendors would support the JMS API.

The JMS API can be broken down into three main parts: the general API, the point-to-point API, and the publish-and-subscribe API. In JMS 1.1, the general API can be used to send and receive messages from either a queue or a topic. The point-to-point API is used solely for messaging with queues, and the publish-and-subscribe API is used solely for messaging using topics.

Within the JMS general API, there are seven main JMS API interfaces related to sending and receiving JMS messages:

- `ConnectionFactory`
- `Destination`
- `Connection`
- `Session`
- `Message`
- `MessageProducer`
- `MessageConsumer`

Of these general interfaces, the `ConnectionFactory` and `Destination` must be obtained from the provider using JNDI (per the JMS specification). The other interfaces are created through factory methods in the various API interfaces. For example, once you have a `ConnectionFactory`, you can create a `Connection`. Once you have a `Connection`, you can create a `Session`. Once you have a `Session`, you can create a `Message`, `MessageProducer`, and `MessageReceiver`. The relationship between these seven primary JMS general API interfaces is illustrated in Figure 1-5.

In JMS, the `Session` object holds the transactional unit of work for messaging, not the `Connection` object. This is different from JDBC, where the `Connection` object holds the transactional unit of work. This means that when using JMS, an application will typically have only a single `Connection` object but will have a pool of `Session` objects.

There are several other interfaces related to exception handling, message priority, and message persistence. These and other API interfaces are discussed in more detail throughout the book and also in Appendix A.

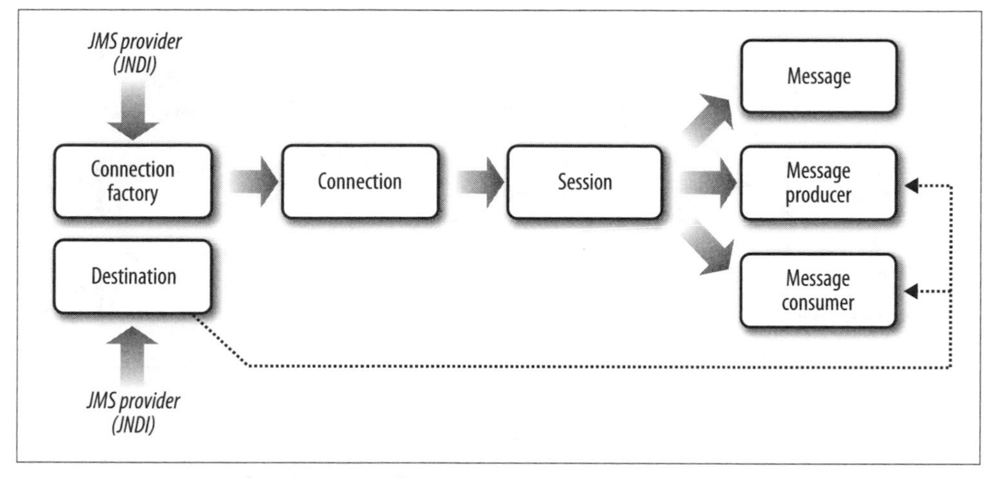

Figure 1-5. JMS general API core interfaces

Point-to-Point API

Once you gain an understanding of the JMS general API, the rest of the JMS API is fairly easy to infer and understand. The point-to-point messaging API refers specifically to the queue-based interfaces within the JMS API. The interfaces used for sending and receiving messages from a queue are as follows:

- `QueueConnectionFactory`
- `Queue`
- `QueueConnection`
- `QueueSession`
- `Message`
- `QueueSender`
- `QueueReceiver`

As in the JMS general API, the `QueueConnectionFactory` and `Queue` objects must be obtained from the JMS provider via JNDI (per the JMS specification). Notice that most of the interface names simply add the word `Queue` before the general API interface name. The exception to this is the `Destination` interface, which is named `Queue`, and the `MessageProducer` and `MessageConsumer` interfaces, which are named `QueueSender` and `QueueReceiver`, respectively. Figure 1-6 illustrates the flow and relationship between the queue-based JMS API interfaces.

Applications using the point-to-point messaging model will typically use the queue-based API rather than the general API.

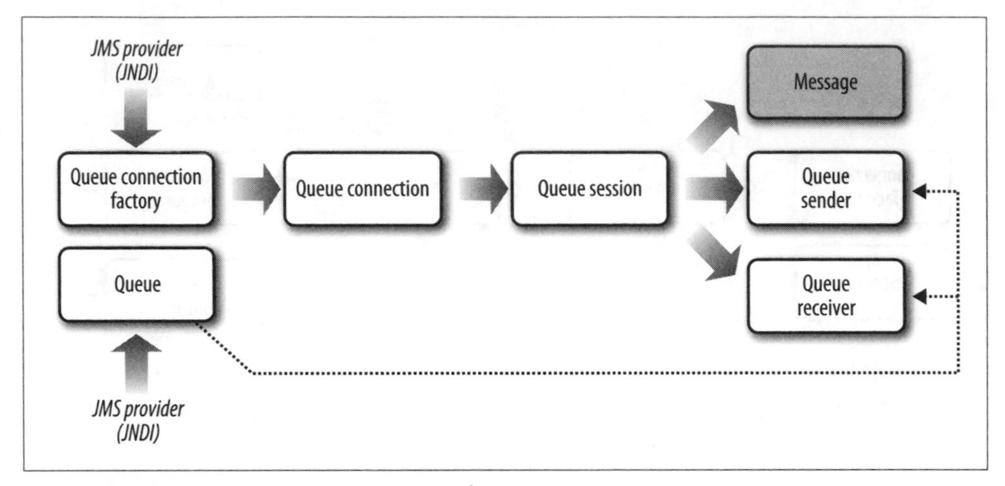

Figure 1-6. JMS point-to-point API core interfaces

Publish-and-Subscribe API

The topic-based JMS API is similar to the queue-based API in that, in most cases, the word `Queue` is replaced with the word `Topic`. The interfaces used within the pub/sub messaging model are as follows:

- `TopicConnectionFactory`
- `Topic`
- `TopicConnection`
- `TopicSession`
- `Message`
- `TopicPublisher`
- `TopicSubscriber`

Notice that the interfaces in the pub/sub domain have names similar to those of the p2p domain, with the exception of `TopicPublisher` and `TopicSubscriber`. The JMS API is very intuitive in this regard. As stated at the start of this chapter, pub/sub uses *topics* with *publishers* and *subscribers*, whereas p2p uses *queues* with *senders* and *receivers*. Notice how this terminology matches the API interface names. The relationship and flow of the topic-based JMS API interfaces are illustrated in Figure 1-7.

Real-World Scenarios

Until now, our discussion of enterprise messaging has been somewhat abstract. This section attempts to give some real-world scenarios to provide you with a better idea of the types of problems that enterprise messaging systems can solve

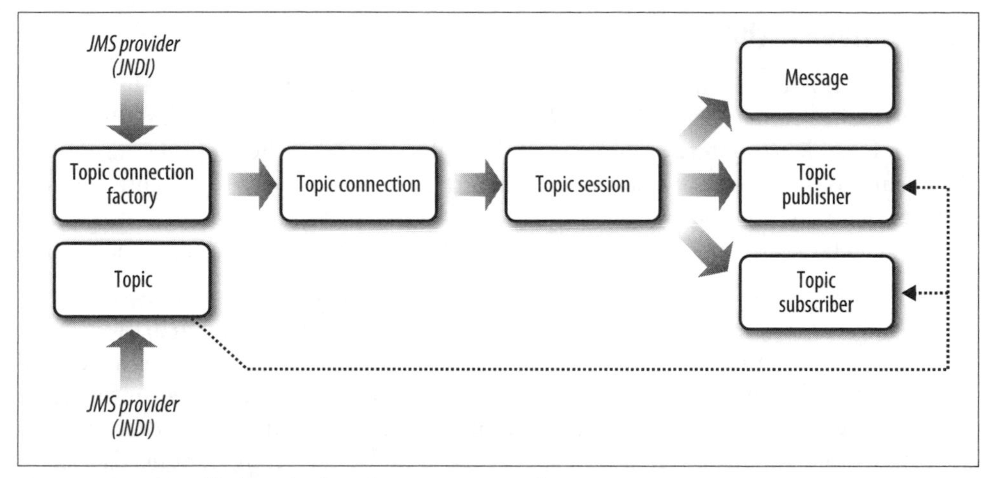

Figure 1-7. JMS publish-and-subscribe API core interfaces

Service-Oriented Architecture

Service-Oriented Architecture (SOA) describes an architecture style that defines business services that are abstracted from the corresponding enterprise service implementations. SOA has given rise to a new breed of middleware known as an Enterprise Service Bus, or ESB. In the early days of SOA, most ESBs were implemented as message brokers, whereby components within the messaging layer were used to perform some sort of intelligent routing or message transformation before delivering the message. These earlier message brokers have evolved into sophisticated commercial and open source ESB products that use messaging at their core. Although some ESB products support a traditional non-JMS HTTP transport, most enterprise-wide production implementations still leverage messaging as the protocol for communication.

Messaging is an excellent means of building the abstraction layer within SOA needed to fully abstract a business service from its underlying implementation. Through messaging, the business service does not need to be concerned about where the corresponding implementation service is, what language it is written in, what platform it is deployed in, or even the name of the implementation service. Messaging also provides the scalability needed within an SOA environment, and also provides a robust level of monitoring and control for requests coming into and out of an ESB. Almost all of the commercial and open source ESB products available today support JMS messaging as a communication protocol—the notable exception being the Microsoft line of messaging products (e.g., BizTalk and MSMQ).

The increased interest and use of SOA in the industry has in turn given rise to increased interest and usage of messaging solutions in general. Although full-blown SOA implementations are continuing to evolve, many companies are tuning to messaging solutions as a step toward SOA.

Event-Driven Architecture

Event-Driven Architecture (EDA) is an architecture style that is built on the premise that the orchestration of processes and events is dynamic and very complex, and therefore not feasible to control or implement through a central orchestration component. When an action takes place in a system, that process sends an event to the entire system stating that an action took place (an event). That event may then kick off other processes, which in turn may kick off additional processes, all decoupled from each other.

Some good examples of EDA include the insurance domain and the defined benefits domain. Both of these industry domains are driven by events that happen in the system. For example, something as simple as changing your address can affect many aspects of the insurance domain, including policies, quotes, and customer records. In this case, the driving event in the insurance application is an address change. However, it is not the responsibility of the address change module to know everything that needs to happen as a result of that event. Therefore, the address change module sends an event message letting the system know that an address has changed. The quoting system will pick up that event and adjust any outstanding quotes that may be present for that customer. Simultaneously, the policy system will pick up the change address event and adjust the rates and policies for that customer.

Another example of EDA is within the defined benefits domain. Getting married or changing jobs triggers events in the system that qualify you for certain changes to your health and retirement benefits. Many of these systems use EDA to avoid using a large, complex, and unmaintainable central processing engine to control all of the actions associated with a particular "qualifying event."

Messaging is the foundation for systems based on an Event-Driven Architecture. Events are typically implemented as empty payload messages containing some information about the event in the header of the message, although some pass the application data as part of the event. Not surprisingly, most architectures based on EDA leverage the pub/sub model as a means of broadcasting the events within a system.

Heterogeneous Platform Integration

Most companies, through a combination of mergers, acquisitions, migrations, or bad decisions, have a myriad of heterogeneous platforms, products, and languages supporting the business. Integrating these platforms can be a challenging task, particularly with standards continually changing and evolving. Messaging plays a key role in being able to make these heterogeneous platforms communicate with one another, whether it be Java EE and Microsoft .NET, Java EE and CICS, or Java EE and Tuxedo C++.

Although platforms such as Java can utilize the JMS API, other platforms such as .NET or C++ cannot (for obvious reasons). Many messaging vendors, both commercial and open source, support both the JMS API and a native API. These providers typically have a built-in messaging bridge that allows the provider to be able to convert a JMS message into an internal message and vice versa. Some platforms, such as .NET, may require an external messaging bridge to convert a JMS message into an MSMQ message (depending on the messaging provider you are using). For example, ActiveMQ provides a messaging bridge for converting MSMQ to JMS (and vice versa). This lower-level platform integration has given rise to a broader scope of integration, known as Enterprise Application Integration.

Enterprise Application Integration

Most mature organizations have both legacy and new applications that are implemented independently and cannot interoperate. In many cases, organizations have a strong desire to integrate these applications so that they can share information and cooperate in larger enterprise-wide operations. The integration of these applications is generally called Enterprise Application Integration (EAI).

A variety of vendor and home-grown solutions are used for EAI, but enterprise messaging systems are central to most of them. Enterprise messaging systems allow stovepipe applications (consisting of heterogeneous products, technologies, and components) to communicate events and to exchange data while remaining physically independent. Data and events can be exchanged in the form of messages via topics or queues, which provide an abstraction that decouples participating applications.

As an example, a messaging system might be used to integrate an Internet order processing system with an Enterprise Resource Planning (ERP) system like SAP. The Internet system uses JMS to deliver business data about new orders to a topic. An ERP gateway application, which accesses a SAP application via its native API, can subscribe to the order topic. As new orders are broadcast to the topic, the gateway receives the orders and enters them into the SAP application.

Business-to-Business

Historically, businesses exchanged data using Electronic Data Interchange (EDI) systems. Data was exchanged using rigid, fixed formats over proprietary Value-Added Networks (VANs). Cost of entry was high and data was usually exchanged in batch processes—not as real-time business events.

The Internet, XML, and modern messaging systems have radically changed how businesses exchange data and interact in what is now called Business-to-Business (B2B). The use of messaging systems is central to modern B2B solutions because it allows organizations to cooperate without requiring them to tightly integrate their business systems. In addition, it lowers the barriers to entry since finer-grained participation is possible. Businesses can join in B2B and disengage depending on the queues and topics with which they interact.

A manufacturer, for example, can set up a topic for broadcasting requests for bids on raw materials. Suppliers can subscribe to the topic and respond by producing messages back to the manufacturer's queue. Suppliers can be added and removed at will, and new topics and queues for different types of inventory and raw materials can be used to partition the systems appropriately.

Geographic Dispersion

These days many companies are geographically dispersed. Brick-and-mortar, click-and-mortar, and dot-coms all face problems associated with geographic dispersion of enterprise systems. Inventory systems in remote warehouses need to communicate with centralized back-office ERP systems at corporate headquarters. Sensitive employee data that is administered locally at each subsidiary needs to be synchronized with the main office. JMS messaging systems can ensure the safe and secure exchange of data across a geographically distributed business.

Information Broadcasting

Auction sites, stock quote services, and securities exchanges all have to push data out to huge populations of recipients in a one-to-many fashion. In many cases, the broadcast of information needs to be selectively routed and filtered on a per-recipient basis. While the outgoing information needs to be delivered in a one-to-many fashion, often the response to such information needs to be sent back to the broadcaster. This is another situation in which enterprise messaging is extremely useful, since pub/sub can be used to distribute the messages and p2p can be used for responses.

Choices in reliability of delivery are key in these situations. In the case of broadcasting stock quotes, for example, absolutely guaranteeing the delivery of information may not be critical, since another broadcast of the same ticker symbol will likely happen in another short interval of time. In the case where a trader is responding to a price quote with a buy order, however, it is crucial that the response is returned in a guaranteed fashion. In this case, you mix reliability of messaging so that the pub/sub distribution is fast but unreliable, while the use of p2p for buying orders from traders is very reliable. JMS and enterprise messaging provide these varying degrees of reliability for both the pub/sub and p2p models.

Building Dynamic Systems

In JMS, pub/sub topics and p2p queues are centrally administered and are referred to as JMS *administered objects*. Your application does not need to know the network location of topics or queues to communicate with other applications; it just uses topic and queue objects as identifiers. Using topics and queues provides JMS applications with a certain level of location transparency and flexibility that makes it possible to add and remove participants in an enterprise system.

For example, a system administrator can dynamically add subscribers to specific topics on an as-needed basis. A common scenario might be if you discover a need to add an audit-trail mechanism for certain messages and not others. Figure 1-8 shows you how to plug in a specialized auditing and logging JMS client whose only job is to track specific messages, just by subscribing to the topics you are interested in.

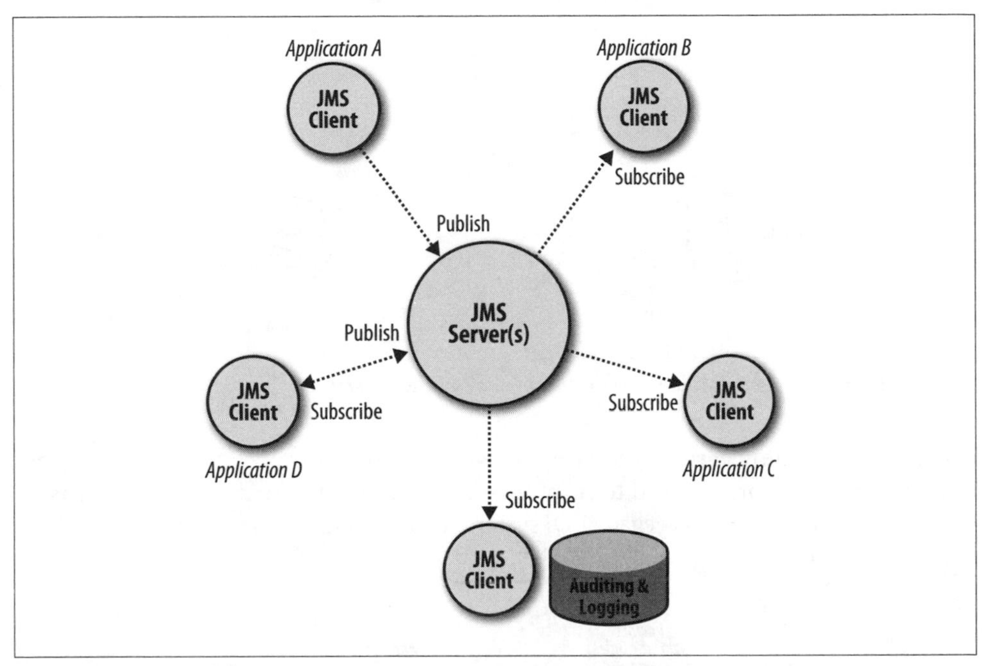

Figure 1-8. Dynamically adding auditing and logging using publish-and-subscribe

The ability to add and remove producers and consumers allows enterprise systems to dynamically alter the routing and re-routing of messages in an already deployed environment.

As another example, we can build on the EAI scenario discussed previously. In this case, a gateway accepts incoming purchase orders, converts them to the format appropriate for a legacy ERP system, and calls into the ERP system for processing (see Figure 1-9).

In Figure 1-8, other JMS applications (A and B) also subscribe to the purchase order topic and do their own independent processing. Application A might be a legacy application in the company, while application B may be another company's business system, representing a B2B integration.

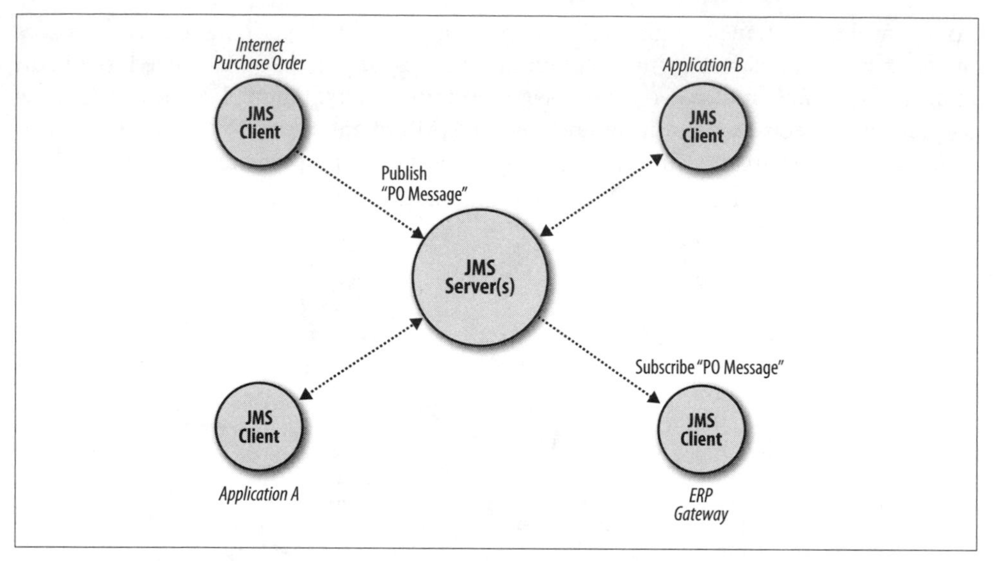

Figure 1-9. Integrating a purchase order system with an ERP system

Using JMS, it's fairly easy to add and remove applications from this process. For example, if purchase orders need to be processed from two different sources, such as an Internet-based system and a legacy EDI system, you can simply add the legacy purchase order system to the mix (see Figure 1-10).

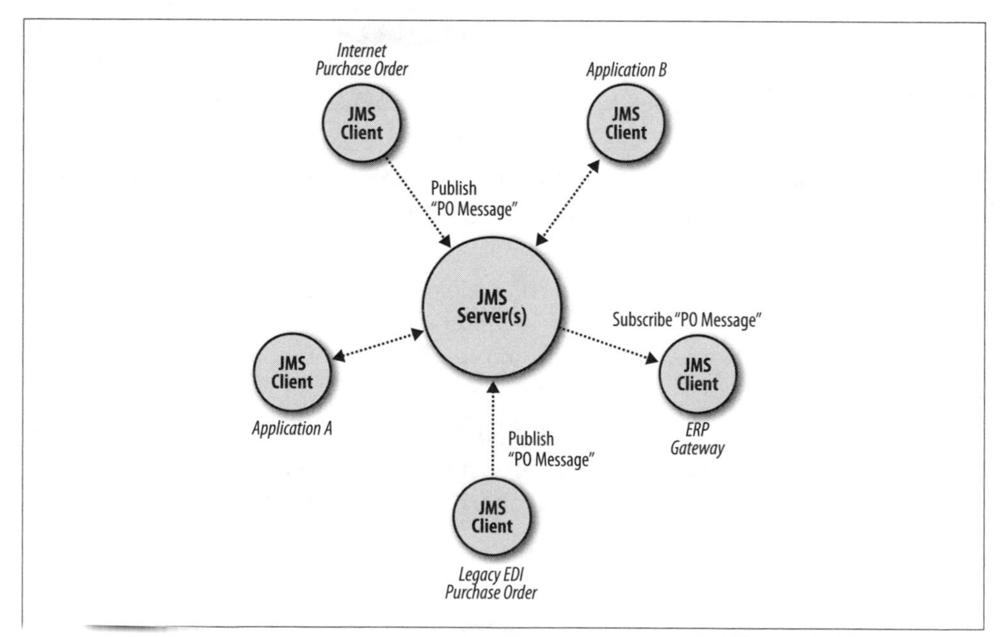

Figure 1-10. Integrating two different purchase order systems with an ERP system

What is interesting about this example is that the ERP gateway is unaware that it is receiving purchase order messages from two completely different sources. The legacy EDI system may be an older in-house system or it could be the main system for a business partner or a recently acquired subsidiary. In addition, the legacy EDI system would have been added dynamically without requiring the shutdown and retooling of the entire system. Enterprise messaging systems make this kind of flexibility possible, and JMS allows Java clients to access many different messaging systems using the same Java programming model.

RPC Versus Asynchronous Messaging

RPC (Remote Procedure Call) is a term commonly used to describe a distributed computing model that is used today by both the Java and .NET platforms. Component-based architectures such as Enterprise JavaBeans are built on top of this model. RPC-based technologies have been, and will continue to be, a viable solution for many applications. However, the enterprise messaging model is superior in certain types of distributed applications. In this section we will discuss the pros and cons of each model.

Tightly Coupled RPC

One of the most successful areas of the tightly coupled RPC model has been in building 3-tier, or *n* -tier, applications. In this model, a presentation layer (first tier) communicates using RPC with business logic on the middle tier (second tier), which accesses data housed on the backend (third tier). Sun Microsystems' J2EE platform and Microsoft's .NET platform are the most modern examples of this architecture.

With J2EE, JSP and servlets represent the presentation tier while Enterprise JavaBeans (EJB) is the middle tier. Regardless of the platform, the core technology used in these systems is RPC-based middleware with RPC being the defining communication paradigm.

RPC attempts to mimic the behavior of a system that runs in one process. When a remote procedure is invoked, the caller is blocked until the procedure completes and returns control to the caller. This synchronized model allows the developer to view the system as if it runs in one process. Work is performed sequentially, ensuring that tasks are completed in a predefined order. The synchronized nature of RPC tightly couples the client (the software making the call) to the server (the software servicing the call). The client cannot proceed—it is blocked—until the server responds.

The tightly coupled nature of RPC creates highly interdependent systems where a failure on one system has an immediate and debilitating impact on other systems. In J2EE, for example, the EJB server must be functioning properly if the servlets that use enterprise beans are expected to function.

RPC works well in many scenarios, but its synchronous, tightly coupled nature is a severe handicap in system-to-system processing where vertical applications are integrated together. In system-to-system scenarios, the lines of communication between vertical systems are many and multidirectional, as Figure 1-11 illustrates.

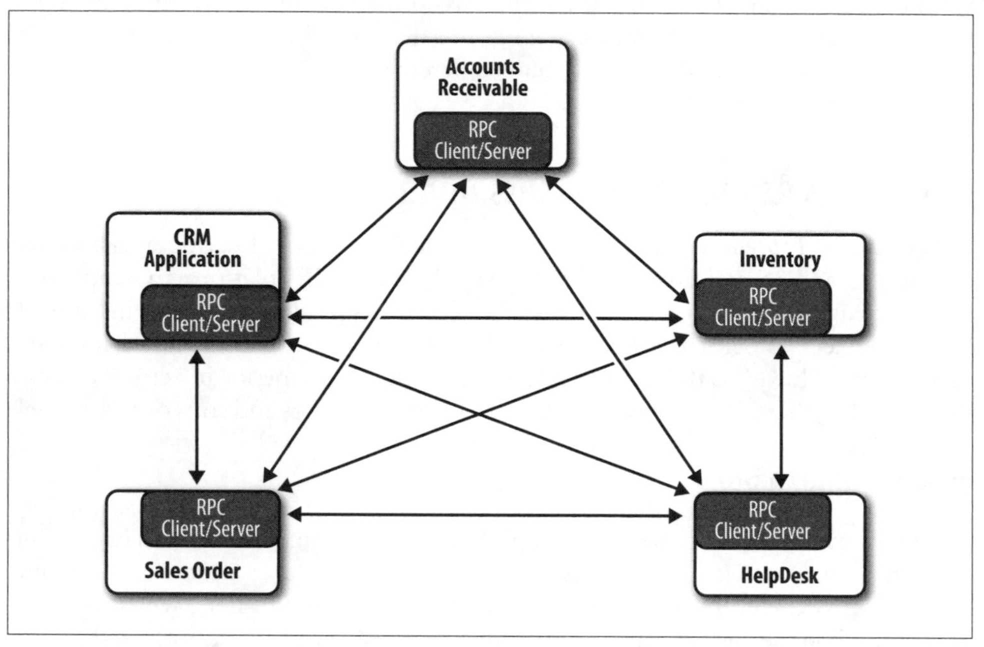

Figure 1-11. Tightly coupled with synchronous RPC

Consider the challenge of implementing this infrastructure using a tightly coupled RPC mechanism. There is the many-to-many problem of managing the connections between these systems. When you add another application to the mix, you have to go back and let all the other systems know about it. Also, systems can crash. Scheduled downtimes need to happen. Object interfaces need to be upgraded.

When one part of the system goes down, everything halts. When you post an order to an order entry system, it needs to make a synchronous call to each of the other systems. This causes the order entry system to block and wait until each system is finished processing the order.*

It is the synchronized, tightly coupled, interdependent nature of RPC systems that cause entire systems to fail as a result of failures in subsystems. When the tightly coupled nature of RPC is not appropriate, as in system-to-system scenarios, messaging provides an alternative.

* Multithreading and looser RPC mechanisms like CORBA's one-way call are options, but these solutions have their own complexities and require very sophisticated development. Threads are expensive when not used wisely, and CORBA one-way calls still require application-level error handling for failure conditions.

Enterprise Messaging

Problems with the availability of subsystems are not an issue with Message-Oriented Middleware. A fundamental concept of messaging is that communication between applications is intended to be asynchronous. Code that is written to connect the pieces together assumes there is a *one-way* message that requires no immediate response from another application. In other words, there is no blocking. Once a message is sent, the messaging client can move on to other tasks; it doesn't have to wait for a response. This is the major difference between RPC and asynchronous messaging, and it is critical to understanding the advantages offered by messaging systems.

In an asynchronous messaging system, each subsystem (Accounts Receivable, Inventory, etc.) is decoupled from the other systems (see Figure 1-12). They communicate through the messaging server, so that a failure in one does not impede the operation of the others.

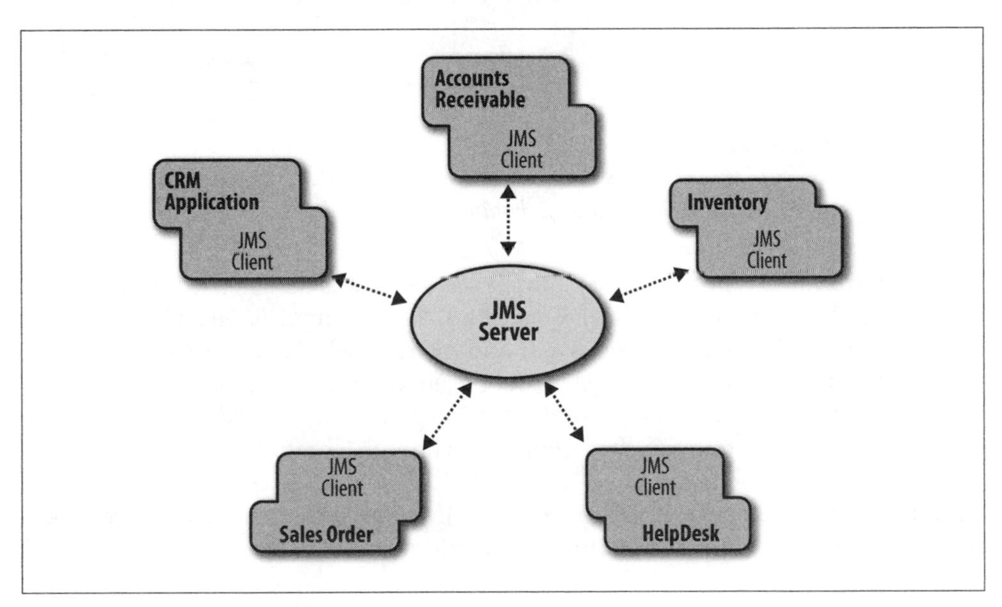

Figure 1-12. JMS provides a loosely coupled environment where partial failure of system components does not impede overall system availability

Partial failure in a networked system is a fact of life. One of the systems may have an unpredictable failure or may need to be shut down at some time during its continuous operation. This can be further magnified by geographic dispersion of in-house and partner systems. In recognition of this, JMS provides *guaranteed delivery*, which ensures that intended consumers will eventually receive a message even if partial failure occurs.

Guaranteed delivery uses a *store-and-forward* mechanism, which means that the underlying message server will write the incoming messages out to a persistent store if the intended consumers are not currently available. When the receiving applications become available at a later time, the store-and-forward mechanism will deliver all of the messages that the consumers missed while unavailable (see Figure 1-13).

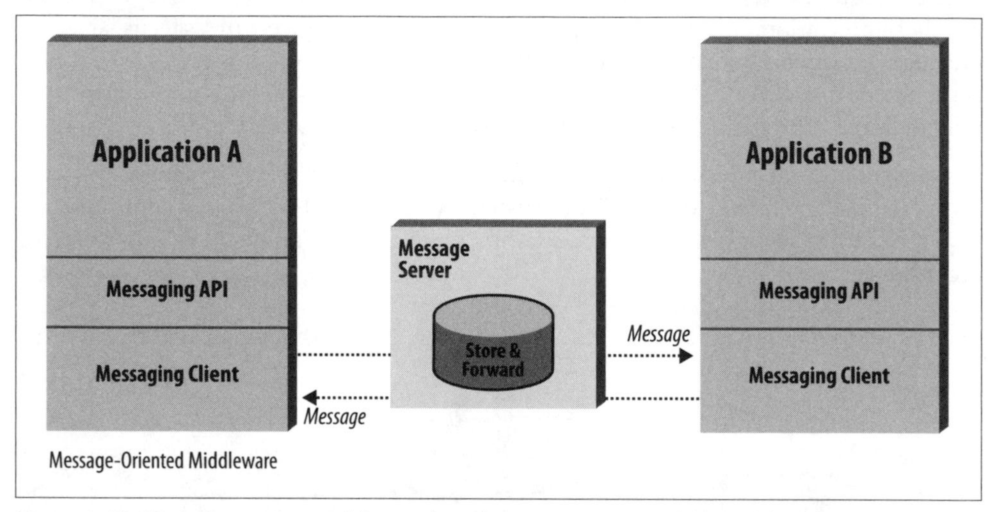

Figure 1-13. Underlying store-and-forward mechanisms guarantee delivery of messages

To summarize, JMS is not just another event service. It was designed to cover a broad range of enterprise applications, including EAI, B2B, push models, etc. Through asynchronous processing, store-and-forward, and guaranteed delivery, it provides high availability capabilities to keep business applications in continuous operation with uninterrupted service. It offers flexibility of integration by providing publish-and-subscribe and point-to-point functionality. Through location transparency and administrative control, it allows for a robust, service-based architecture. And most important, it is extremely easy to learn and use. In the next chapter we will take a look at how simple it is by building our first JMS application.

CHAPTER 2

Developing a Simple Example

Now that you understand Message-Oriented Middleware and some JMS concepts, you are ready to write your first JMS application. Although it would be easier to demonstrate a simple example using the more widely used point-to-point model, the publish-and-subscribe model offers a more interesting example. Therefore, in this chapter we will provide a gentle introduction to JMS using the publish-and-subscribe messaging model. You will get your feet wet with JMS and learn some of the basic classes and interfaces. Chapter 4 covers the point-to-point model in detail using a real-world example, and Chapter 5 covers the publish-and-subscribe messaging model following up on the same example.

As with all examples in this book, example code and instructions specific to several vendors are available for download on O'Reilly's website at *http://oreilly.com/catalog/9780596522049/examples*. You will need to install and configure your JMS provider according to the instructions provided by your vendor. To illustrate a sample vendor configuration, we will be using ActiveMQ version 5.2, a popular robust and production-quality open source JMS provider (see *http://activemq.apache.org*). You can find the basic installation instructions and configuration settings for executing the code examples in this book in Appendix D.

The Chat Application

Internet chat provides an interesting application for learning about the JMS pub/sub messaging model. Used mostly for entertainment, web-based chat applications can be found on thousands of websites. In this type of application, people join virtual chat rooms where they can "chat" with a group of other people.

To illustrate how JMS works, we will use the JMS pub/sub API to build a simple chat application. The requirements of Internet chat map neatly onto the publish-and-subscribe messaging model. In this model, a producer can send a message to many consumers by delivering the message to a single topic. A message producer is also called a *publisher*, and a message consumer is also called a *subscriber*.

The following complete source code listing is a JMS-based chat client. Every participant in a chat session uses this Chat program to join a specific chat room (topic), and to deliver and receive messages to and from that room. In this chapter, we will be taking this example apart and explaining the various API calls used throughout this listing. Further details will be provided in the chapters describing the point-to-point (Chapter 4) and publish-and-subscribe (Chapter 5) models:

```
package ch02.chat;

import java.io.*;
import javax.jms.*;
import javax.naming.*;

public class Chat implements javax.jms.MessageListener {
    private TopicSession pubSession;
    private TopicPublisher publisher;
    private TopicConnection connection;
    private String username;

    /* Constructor used to Initialize Chat */
    public Chat(String topicFactory, String topicName, String username)
        throws Exception {

        // Obtain a JNDI connection using the jndi.properties file
        InitialContext ctx = new InitialContext();

        // Look up a JMS connection factory and create the connection
        TopicConnectionFactory conFactory =
            (TopicConnectionFactory)ctx.lookup(topicFactory);
        TopicConnection connection = conFactory.createTopicConnection();

        // Create two JMS session objects
        TopicSession pubSession = connection.createTopicSession(
            false, Session.AUTO_ACKNOWLEDGE);
        TopicSession subSession = connection.createTopicSession(
            false, Session.AUTO_ACKNOWLEDGE);

        // Look up a JMS topic
        Topic chatTopic = (Topic)ctx.lookup(topicName);

        // Create a JMS publisher and subscriber. The additional parameters
        // on the createSubscriber are a message selector (null) and a true
        // value for the noLocal flag indicating that messages produced from
        // this publisher should not be consumed by this publisher.
        TopicPublisher publisher =
            pubSession.createPublisher(chatTopic);
        TopicSubscriber subscriber =
            subSession.createSubscriber(chatTopic, null, true);

        // Set a JMS message listener
        subscriber.setMessageListener(this);

        // Intialize the Chat application variables
        this.connection = connection;
```

```
        this.pubSession = pubSession;
        this.publisher = publisher;
        this.username = username;

        // Start the JMS connection; allows messages to be delivered
        connection.start();
    }


    /* Receive Messages From Topic Subscriber */
    public void onMessage(Message message) {
        try {
            TextMessage textMessage = (TextMessage) message;
            System.out.println(textMessage.getText());
        } catch (JMSException jmse){ jmse.printStackTrace(); }
    }

    /* Create and Send Message Using Publisher */
    protected void writeMessage(String text) throws JMSException {
        TextMessage message = pubSession.createTextMessage();
        message.setText(username+": "+text);
        publisher.publish(message);
    }

    /* Close the JMS Connection */
    public void close() throws JMSException {
        connection.close();
    }

    /* Run the Chat Client */
    public static void main(String [] args) {
        try {
            if (args.length!=3)
                System.out.println("Factory, Topic, or username missing");

            // args[0]=topicFactory; args[1]=topicName; args[2]=username
            Chat chat = new Chat(args[0],args[1],args[2]);

            // Read from command line
            BufferedReader commandLine = new
              java.io.BufferedReader(new InputStreamReader(System.in));

            // Loop until the word "exit" is typed
            while(true) {
                String s = commandLine.readLine();
                if (s.equalsIgnoreCase("exit")){
                    chat.close();
                    System.exit(0);
                } else
                    chat.writeMessage(s);
            }
        } catch (Exception e) { e.printStackTrace(); }
    }
}
```

Notice that the code just given is using the form of the createSubscriber() method that takes three arguments rather than just one. This is so the noLocal flag can be set (the third parameter) so that messages published by this class will not also be consumed by this class. The second parameter is used for a message selector. Since we are not doing any filtering on the topic, this value is set to null. If we were to use the single parameter method call to create a subscriber, we would see the messages we sent on our console display.

Getting Started with the Chat Example

To run the Chat application you will need a JMS provider that supports JNDI and JMS 1.1. To illustrate some of the details and configuration in the code examples, we'll be using ActiveMQ, a popular open source JMS provider. You'll need to consult your JMS vendor's documentation for information on configuring a TopicConnectionFactory and a Topic for the Chat application. In our example, we have named these TopicCF and topic1, respectively. For instance, using ActiveMQ you can set the TopicConnectionFactory name and a Topic for the Chat Application by creating a *jndi.properties* file located in your classpath and setting the names as follows:

```
java.naming.factory.initial = org.apache.activemq.jndi.ActiveMQInitialContextFactory
java.naming.provider.url = tcp://localhost:61616
java.naming.security.principal=system
java.naming.security.credentials=manager

connectionFactoryNames = TopicCF
topic.topic1 = jms.topic1
```

The *jndi.properties* file also contains the JNDI connection information for the JMS provider. You will need to set the initial context factory class, provider URL, username, and password needed to connect to the JMS server. Each vendor will have a different context factory class and URL name for connecting to the server. You will need to consult the documentation of your specific JMS provider or Java EE container to obtain these values. For example, in the *jndi.properties* file just shown, for ActiveMQ you would set the initial context factory to org.apache.activemq.jndi.ActiveMQInitialContextFactory and the provider URL to tcp://localhost:61616 (the default protocol, host, and port for ActiveMQ). More details surrounding the installation and configuration of ActiveMQ can be found in Appendix D.

After configuring and starting your JMS server, you will need to compile the Chat application. In addition to the *jms-11.jar* file, you will need to include any JAR files required by the JMS provider in your classpath (in the case of ActiveMQ 5.2, simply include the *activemq-all-5.2.0.jar* file in your classpath).

The Chat class includes a main() method so that it can be run as a standalone Java application. You will obviously want to open multiple command windows so that you can simulate a chat with multiple people. The Chat class can be executed from the command line or from a shell script:

```
java ch02.chat.Chat topicConnectionFactory topicName username
```

For example, in the OpenJMS configuration listed previously we have defined a Topic Connection Factory named `TopicCF` and a Topic named `topic1`. Therefore, to execute the chat application for a user named `Fred` and another user named `Wilma`, you would use the following command:

```
java ch02.chat.Chat TopicCF topic1 Fred
java ch02.chat.Chat TopicCF topic1 Wilma
```

Run at least two chat clients in separate command windows and try typing into one; you should see the text you type displayed by the other client.

Before we examine the source code in detail, a quick explanation of what the code is doing might be helpful. The chat client creates a JMS publisher and subscriber for a specific topic. The topic represents the chat room. The JMS server registers all the JMS clients that want to publish or subscribe to a specific topic. When text is entered at the command line of one of the chat clients, it is published to the messaging server. The messaging server identifies the topic associated with the publisher and delivers the message to all the JMS clients that have subscribed to that topic. As Figure 2-1 illustrates, messages published by any one of the JMS clients are delivered to all the JMS subscribers for that topic.

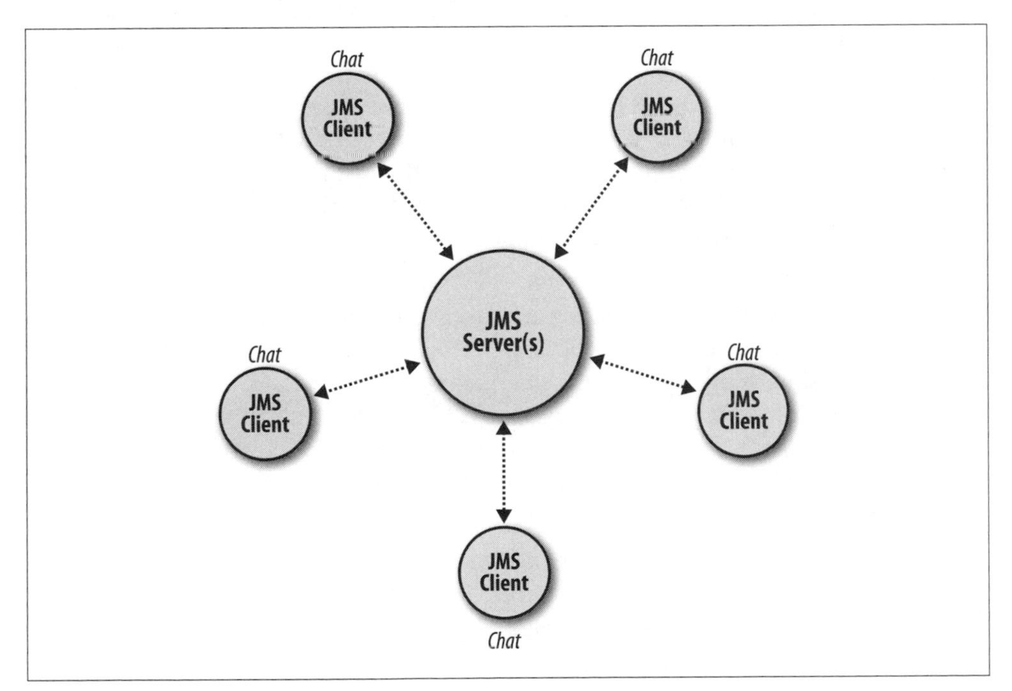

Figure 2-1. The Chat application

Examining the Source Code

Running the `Chat` example in a couple of command windows demonstrates *what* the `Chat` application does. The rest of this chapter examines the source code for the `Chat` application so that you can see *how* the `Chat` application works.

Bootstrapping the JMS client

The `main()` method bootstraps the chat client and provides a command-line interface. Once an instance of the `Chat` class is created, the `main()` method spends the rest of its time reading text typed at the command line and passing it to the `Chat` instance using the instance's `writeMessage()` method.

The `Chat` instance connects to the topic and receives and delivers messages. The `Chat` instance starts its life in the constructor, which does all the work to connect to the topic and set up the `TopicPublisher` and `TopicSubscribers` for delivering and receiving messages.

Obtaining a JNDI connection

The chat client starts by obtaining a JNDI connection to the JMS messaging server. JNDI is an implementation-independent API for directory and naming systems. A directory service provides JMS clients with access to `ConnectionFactory` and `Destination` (topics and queues) objects. `ConnectionFactory` and `Destination` objects are the only things in JMS that cannot be obtained using the JMS API—unlike connections, sessions, producers, consumers, and messages, which are manufactured using the factory pattern within the JMS API. JNDI provides a convenient, location-transparent, configurable, and portable mechanism for obtaining `ConnectionFactory` and `Destination` objects, also called JMS-administered objects because they are established and configured by a system administrator.

Using JNDI, a JMS client can obtain access to a JMS provider by first looking up a `ConnectionFactory`. The `ConnectionFactory` is used to create JMS connections, which can then be used for sending and receiving messages. `Destination` objects, which represent virtual channels (topics and queues) in JMS, are also obtained via JNDI and are used by the JMS client. The directory service can be configured by the system administrator to provide JMS-administered objects so that the JMS clients don't need to use proprietary code to access a JMS provider.

JMS servers will either work with a separate directory service (e.g., LDAP) or provide their own directory service that supports the JNDI API. For more details on JNDI, see the sidebar "Understanding JNDI" on page 31.

The constructor of the `Chat` class starts by obtaining a connection to the JNDI naming service used by the JMS server:

```
// Obtain a JNDI connection using the jndi.properties file
InitialContext ctx = new InitialContext();
```

Understanding JNDI

JNDI is a standard Java extension that provides a uniform API for accessing a variety of directory and naming services. In this respect, it is somewhat similar to JDBC. JDBC lets you write code that can access different relational databases such as Oracle, SQLServer, or Sybase; JNDI lets you write code that can access different directory and naming services, such as LDAP, NDS, CORBA Naming Service, and proprietary naming services provided by JMS servers.

In JMS, JNDI is used mostly as a naming service to locate administered objects. Administered objects are JMS objects that are created and configured by the system administrator. Administered objects include JMS `ConnectionFactory` and `Destination` objects such as topics and queues.

Administered objects are bound to a name in a naming service. A naming service associates names with distributed objects, files, and devices so that they can be located on the network using simple names instead of cryptic network addresses. An example of a naming service is the DNS, which converts an Internet hostname like *www.oreilly.com* into a network address that browsers use to connect to web servers. There are many other naming services, such as CosNaming in CORBA and the Java RMI registry. Naming services allow printers, distributed objects, and JMS administered objects to be bound to names and organized in a hierarchy similar to a filesystem. A directory service is a more sophisticated kind of naming service.

JNDI provides an abstraction that hides the specifics of the naming service, making client applications more portable. Using JNDI, JMS clients can browse a naming service and obtain references to administered objects without knowing the details of the naming service or how it is implemented. JMS servers can usually be used in combination with a standard JNDI driver (a.k.a. service provider) and directory service like Lightweight Directory Access Protocol (LDAP), or provide a proprietary JNDI service provider and directory service.

JNDI is both virtual and dynamic. It is virtual because it allows one naming service to be linked to another. Using JNDI, you can drill down through directories to files, printers, JMS administered objects, and other resources following virtual links between naming services. The user doesn't know or care where the directories are actually located. As an administrator, you can create virtual directories that span a variety of different services over many different physical locations.

JNDI is dynamic because it allows the JNDI drivers for specific types of naming and directory services to be loaded dynamically at runtime. A driver maps a specific kind of naming or directory service into the standard JNDI class interfaces. Drivers have been created for LDAP, Novell NetWare NDS, Sun Solaris NIS+, CORBA CosNaming, and many other types of naming and directory services, including proprietary ones. Dynamically loading JNDI drivers (service providers) makes it possible for a client to navigate across arbitrary directory services without knowing in advance what kinds of services it is likely to find.

Creating a connection to a JNDI naming service requires that a `javax.naming.InitialContext` object be created. An `InitialContext` is the starting point for any JNDI lookup—it's similar in concept to the root of a filesystem. The `InitialContext` provides a network connection to the directory service that acts as a root for accessing JMS-administered objects. The properties used to create an `InitialContext` depend on which JMS directory service you are using. You could configure the initial context properties using the `Properties` Object directly in your source code, or preferably using an external *jndi.properties* file located in the classpath of the application. In our example using ActiveMQ, the *jndi.properties* file would look something like this:

```
java.naming.factory.initial = org.apache.activemq.jndi.ActiveMQInitialContextFactory
java.naming.provider.url = tcp://localhost:61616
java.naming.security.principal=system
java.naming.security.credentials=manager

...
```

The corresponding source code using the `properties` object would be as follows:

```
Properties env = new Properties();
env.put(Context.SECURITY_PRINCIPAL, "system");
env.put(Context.SECURITY_CREDENTIALS, "manager");
env.put(Context.INITIAL_CONTEXT_FACTORY,
   "org.apache.activemq.jndi.ActiveMQInitialContextFactory");
env.put(Context.PROVIDER_URL, "tcp://localhost:61616");

InitialContext ctx = new InitialContext(env);
```

The TopicConnectionFactory

Once a JNDI `InitialContext` object is instantiated, it can be used to look up the `TopicConnectionFactory` in the messaging server's naming service:

```
TopicConnectionFactory conFactory =
    (TopicConnectionFactory)ctx.lookup(topicFactory);
```

The `javax.jms.TopicConnectionFactory` is used to manufacture connections to a message server. A `TopicConnectionFactory` is a type of administered object, which means that its attributes and behavior are configured by the system administrator responsible for the messaging server. The `TopicConnectionFactory` is implemented differently by each vendor, so configuration options available to system administrators vary from product to product. A connection factory might, for example, be configured to manufacture connections that use a particular protocol, security scheme, clustering strategy, etc. A system administrator might choose to deploy several different `TopicConnectionFactory` objects, each configured with its own JNDI lookup name.

The `TopicConnectionFactory` provides two overloaded versions of the `createTopicConnection()` method:

```
package javax.jms;

public interface TopicConnectionFactory extends ConnectionFactory {
```

```
    public TopicConnection createTopicConnection()
       throws JMSException, JMSSecurityException;
    public TopicConnection createTopicConnection(String username,
       String password) throws JMSException, JMSSecurityException;
}
```

These methods are used to create `TopicConnection` objects. The behavior of the no-arg method depends on the JMS provider. Some JMS providers will assume that the JMS client is connecting under anonymous security context, whereas other providers may assume that the credentials can be obtained from JNDI or the current thread.* The second method provides the client with a username-password authentication credential, which can be used to authenticate the connection. In our code, we are using the no-arg method, which will use a default user identity when creating the connection.

The TopicConnection

The `TopicConnection` is created by the `TopicConnectionFactory`:

```
// Look up a JMS connection factory and create the connection
TopicConnectionFactory conFactory =
    (TopicConnectionFactory)ctx.lookup(topicFactory);
TopicConnection connection = conFactory.createTopicConnection();
```

The `TopicConnection` represents a connection to the message server. Each `TopicConnection` that is created from a `TopicConnectionFactory` is a unique connection to the server.† A JMS client might choose to create multiple connections from the same connection factory, but this is rare as connections are relatively expensive (each connection requires a network socket, I/O streams, memory, etc.). Creating multiple `session` objects (discussed later in this chapter) from the same connection is considered more efficient, because sessions share access to the same connection. The `TopicConnection` is an interface that extends `javax.jms.Connection` interface. It defines several general-purpose methods used by clients of the `TopicConnection`. Among these methods are the `start()` , `stop()`, and `close()` methods:

```
public interface Connection {
    public void start() throws JMSException;
    public void stop() throws JMSException;
    public void close() throws JMSException;
    ...
}

public interface TopicConnection extends Connection {
    public TopicSession createTopicSession(boolean transacted,
                                           int acknowledgeMode)
    throws JMSException;
```

* Thread-specific storage is used with the Java Authentication and Authorization Service (JAAS) to allow security credentials to transparently propagate between resources and applications.

† The actual physical network connection may or may not be unique, depending on the vendor. However, the connection is considered to be logically unique so authentication and connection control can be managed separately from other connections.

```
    ...
}
```

The `start()`, `stop()`, and `close()` methods allow a client to manage the connection directly. The `start()` method turns the inbound flow of messages "on," allowing messages to be received by the client. This method is used at the end of the constructor in the `Chat` class:

```
// Start the JMS connection; allows messages to be delivered
connection.start();
```

It is a good idea to start the connection *after* the subscribers have been set up, because the messages start to flow in from the topic as soon as `start()` is invoked.

The `stop()` method blocks the flow of inbound messages until the `start()` method is invoked again. The `close()` method is used to close the `TopicConnection` to the message server. This should be done when a client is finished using the `TopicConnection`; closing the connection conserves resources on the client and server. In the `Chat` class, the `main()` method calls `Chat.close()` when "exit" is typed at the command line. The `Chat.close()` method in turn calls the `TopicConnection.close()` method:

```
public void close() throws JMSException {
   connection.close();
}
```

Closing a `TopicConnection` closes all the objects associated with the connection, including the `TopicSession`, `TopicPublisher`, and `TopicSubscriber`.

The TopicSession

After the `TopicConnection` is obtained, it's used to create `TopicSession` objects:

```
// Create two JMS session objects
TopicSession pubSession = connection.createTopicSession(
    false, Session.AUTO_ACKNOWLEDGE);

TopicSession subSession = connection.createTopicSession(
    false, Session.AUTO_ACKNOWLEDGE);
```

A `TopicSession` object is a factory for creating `Message`, `TopicPublisher`, and `TopicSub scriber` objects. It is also used as the transactional unit of work within JMS. A client can create multiple `TopicSession` objects to provide more granular control over publishers, subscribers, and their associated transactions. In this case, we create two `TopicSession` objects: `pubSession` and `subSession`. We need two objects because of threading restrictions in JMS, which are discussed later in this chapter.

The `boolean` parameter in the `createTopicSession()` method indicates whether the `Session` object will be transacted. A transacted `Session` automatically manages outgoing and incoming messages within a transaction. Transactions are important but not critical to our discussion at this time, so the parameter is set to `false`, which means the `TopicSession` will not be transacted. Transactions are discussed in more detail in Chapter 7.

The second parameter indicates the acknowledgment mode used by the JMS client. An acknowledgment is a notification to the message server that the client has received the message. In this case we chose AUTO_ACKNOWLEDGE, which means that the message is automatically acknowledged after it is received by the client.

The TopicSession objects are used to create the TopicPublisher and TopicSubscriber. The TopicPublisher and TopicSubscriber objects are created with a Topic identifier and are dedicated to the TopicSession that created them; they operate under the control of a specific TopicSession:

```
TopicPublisher publisher =
    pubSession.createPublisher(chatTopic);
TopicSubscriber subscriber =
    subSession.createSubscriber(chatTopic);
```

The TopicSession is also used to create the Message objects that are delivered to the topic. The pubSession is used to create Message objects in the writeMessage() method. When you type text at the command line, the main() method reads the text and passes it to the Chat instance by invoking writeMessage(). The writeMessage() method (shown in the following example) uses the pubSession object to generate a TextMessage object that can be used to deliver the text to the topic:

```
protected void writeMessage(String text) throws JMSException {

   TextMessage message = pubSession.createTextMessage();
   message.setText(username+" : "+text);
   publisher.publish(message);
}
```

Several Message types can be created by a TopicSession. The most commonly used type is the TextMessage.

The Topic

JNDI is used to locate a Topic object, which is an administered object like the Topic ConnectionFactory:

```
InitialContext jndi = new InitialContext(env);
...

// Look up a JMS topic
Topic chatTopic = (Topic)jndi.lookup(topicName);
```

A Topic object is a handle or identifier for an actual topic, called a *physical topic*, on the messaging server. A physical topic is an electronic channel to which many clients can subscribe and publish. A topic is analogous to a news group or list server: when a message is sent to a news group or list server, it is delivered to all the subscribers. Similarly, when a JMS client delivers a Message object to a topic, all the clients subscribed to that topic receive the Message.

The Topic object encapsulates a vendor-specific name for identifying a physical topic in the messaging server. The Topic object has one method, getName(), which returns the name identifier for the physical topic it represents. The name encapsulated by a Topic object is vendor-specific and varies from product to product. For example, one vendor might use dot-separated (.) topic names, like "oreilly.jms.chat," while another vendor might use a completely different naming system, similar to LDAP naming, "o=oreilly,cn=chat." Using topic names directly will result in client applications that are not portable across brands of JMS servers. The Topic object hides the topic name from the client, making the client more portable.

As a convention, we'll refer to a physical topic as a *topic* and only use the term "physical topic" when it's important to stress its difference from a Topic object.

The TopicPublisher

A TopicPublisher is created using the pubSession and the chatTopic:

```
// Look up a JMS topic
Topic chatTopic = (Topic)ctx.lookup(topicName);

// Create a JMS publisher and subscriber
TopicPublisher publisher =
    pubSession.createPublisher(chatTopic);
```

A TopicPublisher is used to deliver messages to a specific topic on a message server. The Topic object used in the createPublisher() method identifies the topic that will receive messages from the TopicPublisher. In the Chat example, any text typed on the command line is passed to the Chat class's writeMessage() method. This method uses the TopicPublisher to deliver a message to the topic:

```
/* Create and Send Message Using Publisher */
protected void writeMessage(String text) throws JMSException {
    TextMessage message = pubSession.createTextMessage();
    message.setText(username+": "+text);
    publisher.publish(message);
}
```

The TopicPublisher delivers messages to the topic asynchronously. Asynchronous delivery and consumption of messages is a key characteristic of Message-Oriented Middleware; the TopicPublisher doesn't block or wait until all the subscribers receive the message. Instead, it returns from the publish() method as soon as the message server receives the message. It's up to the message server to deliver the message to all the subscribers for that topic.

The TopicSubscriber

The TopicSubscriber is created using the subSession and the chatTopic:

```
// Look up a JMS topic
Topic chatTopic = (Topic)ctx.lookup(topicName);

// Create a JMS publisher and subscriber
TopicSubscriber subscriber =
    subSession.createSubscriber(chatTopic, null, true);
```

A `TopicSubscriber` receives messages from a specific topic. The `Topic` object argument used in the `createSubscriber()` method identifies the topic from which the `TopicSubscriber` will receive messages. The second argument contains the message selector used to filter out only those messages we want to receive based on certain criteria. In this case, we set this value to `null`, indicating that we want to receive all messages. The third argument contains a boolean value indicating whether or not we want to receive messages we publish ourselves. In this case, we are setting the value to `true`, indicating that, as a subscriber to the topic, we do not want to see messages we publish.

The `TopicSubscriber` receives messages from the message server one at a time (serially). These messages are pushed from the message server to the `TopicSubscriber` asynchronously, which means that the `TopicSubscriber` does not have to poll the message server for messages. In our example, each chat client will receive any message published by any of the other chat clients. When a user enters text at the command line, the text message is delivered to all other chat clients that subscribe to the same topic.

The pub/sub messaging model in JMS includes an in-process Java event model for handling incoming messages. This is similar to the event-driven model used by JavaBeans.‡ An object simply implements the listener interface, in this case the `MessageListener`, and then is registered with the `TopicSubscriber`. A `TopicSubscriber` may have only one `MessageListener` object. Here is the definition of the `MessageListener` interface used in JMS:

```
package javax.jms;

public interface MessageListener {
    public void onMessage(Message message);
}
```

When the `TopicSubscriber` receives a message from its topic, it invokes the `onMessage()` method of its `MessageListener` objects. The `Chat` class itself implements the `MessageListener` interface and implements the `onMessage()` method:

```
public class Chat implements javax.jms.MessageListener {
    ...
    public void onMessage(Message message) {
        try{
            TextMessage textMessage = (TextMessage)message;
            System.out.println(textMessage.getText());
        } catch (JMSException jmse) { jmse.printStackTrace(); }
    }
    ...
}
```

The `Chat` class is a `MessageListener` type, and therefore registers itself with the `TopicSubscriber` in its constructor:

‡ Although the in-process event model used by `TopicSubscriber` is similar to the one used in JavaBeans, JMS itself is an API and the interfaces it defines are not JavaBeans.

```
TopicSubscriber subscriber = subSession.createSubscriber(chatTopic);

subscriber.setMessageListener(this);
```

When the message server pushes a message to the `TopicSubscriber`, the `TopicSubscriber` invokes the `Chat` object's `onMessage()` method.

It's fairly easy to confuse the Java Message Service with its use of a Java event model. JMS is an API for asynchronous distributed enterprise messaging that spans processes and machines across a network. The Java event model is used to deliver events by invoking methods on one or more objects in the same process that have registered as listeners. The JMS pub/sub model uses the Java event model so that a `TopicSubscriber` can notify its `MessageListener` object in the same process that a message has arrived from the message server.

The Message

In the `chat` example, the `TextMessage` class is used to encapsulate the messages we send and receive. A `TextMessage` contains a `java.lang.String` as its body and is the most commonly used message type. The `onMessage()` method receives `TextMessage` objects from the `TopicSubscriber`. Likewise, the `writeMessage()` method creates and publishes `TextMessage` objects using the `TopicPublisher`:

```
/* Receive Messages From Topic Subscriber */
public void onMessage(Message message) {
   try {
      TextMessage textMessage = (TextMessage) message;
      String text = textMessage.getText();
      System.out.println(text);
    } catch (JMSException jmse){ jmse.printStackTrace(); }
}

/* Create and Send Message Using Publisher */
protected void writeMessage(String text) throws JMSException {
   TextMessage message = pubSession.createTextMessage();
   message.setText(username+": "+text);
   publisher.publish(message);
}
```

A message has three parts: a *header*, *properties*, and *payload*. The header is comprised of special fields that are used to identify the message, declare attributes of the message, and provide information for routing. The properties area of the message contains additional metadata about the message that is set by the application developer or, in some cases, the JMS provider (more on this in Chapter 3). The difference between message types is determined largely by their payload (i.e., the type of application data the message contains). The `Message` class, which is the superclass of all message objects, has no payload. It is a lightweight message that delivers no payload but can serve as a simple event notification. The other message types have special payloads that determine their type and use:

`Message`
: This type has no payload. It is useful for simple event notification.

`TextMessage`
: This type carries a `java.lang.String` as its payload. It is useful for exchanging simple text messages and also for more complex character data, such as XML documents.

`ObjectMessage`
: This type carries a serializable Java object as its payload. It's useful for exchanging Java objects.

`BytesMessage`
: This type carries an array of primitive bytes as its payload. It's useful for exchanging data in an application's native format, which may not be compatible with other existing `Message` types. It is also useful where JMS is used purely as a transport between two systems, and the message payload is opaque to the JMS client.

`StreamMessage`
: This type carries a stream of primitive Java types (`int`, `double`, `char`, etc.) as its payload. It provides a set of convenience methods for mapping a formatted stream of bytes to Java primitives. It's an easy programming model when exchanging primitive application data in a fixed order.

`MapMessage`
: This type carries a set of name-value pairs as its payload. The payload is similar to a `java.util.Properties` object, except the values must be Java primitives or their wrappers. The `MapMessage` is useful for delivering keyed data.

Sessions and Threading

The `Chat` application uses a separate session for the publisher and subscriber, `pubSession` and `subSession`, respectively. This is due to a threading restriction imposed by JMS. According to the JMS specification, a session may not be operated on by more than one thread at a time. In our example, two threads of control are active: the default main thread of the `Chat` application and the thread that invokes the `onMessage()` handler. The thread that invokes the `onMessage()` handler is owned by the JMS provider. Since the invocation of the `onMessage()` handler is asynchronous, it could be called while the main thread is publishing a message in the `writeMessage()` method. If both the publisher and subscriber had been created by the same session, the two threads could operate on these methods at the same time; in effect, they could operate on the same `TopicSession` concurrently—a condition that is prohibited.

A goal of the JMS specification was to avoid imposing an internal architecture on the JMS provider. Requiring a JMS provider's implementation of a `Session` object to be capable of safely handling multiple threads was specifically avoided. This is mostly due to one of the intended uses of JMS—that the JMS API be a wrapper around an existing messaging system, which may not have multithreaded delivery capabilities on the client.

The requirement imposed on the JMS provider is that the sending of messages and the asynchronous receiving of messages be processed serially. It is possible to publish-and-subscribe using the same session, but only if the application is publishing from within the `onMessage()` handler. An example of this will be covered in Chapter 5.

CHAPTER 3

Anatomy of a JMS Message

This chapter focuses on the anatomy of a message: the individual parts that make up a message (headers, properties, and the different kinds of message payloads). Appendixes B and C cover additional information that will prove invaluable as a reference when developing JMS applications. Appendix B provides in-depth information on the purpose and application of JMS headers, and Appendix C covers the rules governing the use of JMS properties. Although you do not need to read these appendixes to understand subsequent chapters in this book, you will need them as a reference when implementing real JMS applications. After you finish reading this chapter, take a look at Appendixes B and C so you're familiar with their content.

The `Message` is the most important part of the entire JMS specification. All data and events in a JMS application are communicated with messages, while the rest of JMS exists to facilitate the transfer of messages. Messages are the lifeblood of the system.

A JMS message carries application data and provides event notification. Its role is unique to distributed computing. In RPC-based systems (CORBA, Java RMI, DCOM), a message is a command to execute a method or procedure, which blocks the sender until a reply has been received. A JMS message is not a command; it transfers data and tells the receiver that something has happened. A message doesn't dictate what the recipient should do and the sender doesn't wait for a response. This decouples the sender from the receiver, making messaging systems and their messages far more dynamic and flexible than request/reply paradigms.

A `Message` object has three parts: the message header, message properties, and finally the message data itself, called the payload or message body (see Figure 3-1).

Messages come in various types that are defined by the payload they carry. The payload itself might be very structured, as with `StreamMessage` and `BytesMessage` objects, or fairly unstructured, as with `TextMessage`, `ObjectMessage`, and `MapMessage` types. Messages can carry important data or simply serve as notifications of events in the system. In most cases, messages are both notifications and vehicles for carrying data.

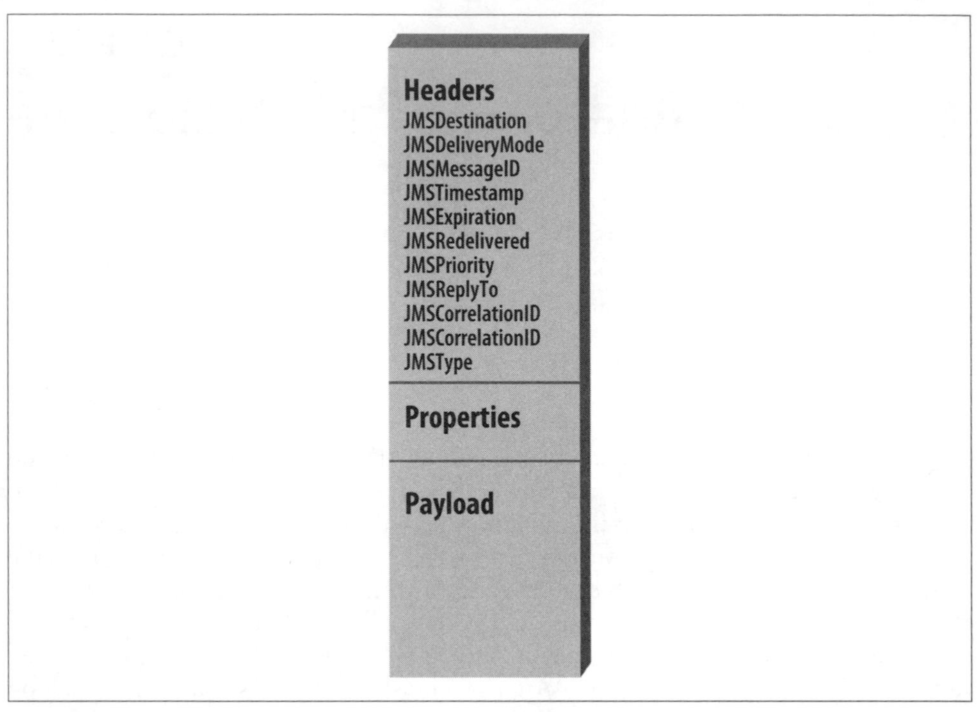

Figure 3-1. Anatomy of a message

The `message` headers provide metadata about the message describing who or what created the message, when it was created, how long the data is valid, etc. The headers also contain routing information that describes the destination of the message (topic or queue), how a message should be acknowledged, and a lot more. In addition to headers, messages can carry properties that can be defined and set by the JMS client. JMS consumers can choose to receive messages based on the values of certain headers and properties, using a special filtering mechanism called *message selectors*. Message selectors are discussed in detail in Chapter 6.

Headers

Every JMS message has a set of standard headers. Each header is identified by a set of accessor and mutator methods that follow the idiom **setJMS<HEADER>()**, **getJMS<HEADER>()**. Here is a partial definition of the **Message** interface that shows all the JMS header methods:

```
public interface Message {

    public Destination getJMSDestination() throws JMSException;
    public void setJMSDestination(Destination destination)
    throws JMSException;
```

```
    public int getJMSDeliveryMode() throws JMSException;
    public void setJMSDeliveryMode(int deliveryMode)
    throws JMSException;

    public String getJMSMessageID() throws JMSException;
    public void setJMSMessageID(String id) throws JMSException;

    public long getJMSTimestamp() throws JMSException;
    public void setJMSTimestamp(long timestamp) throws JMSException;

    public long getJMSExpiration() throws JMSException;
    public void setJMSExpiration(long expiration) throws JMSException;

    public boolean getJMSRedelivered() throws JMSException;
    public void setJMSRedelivered(boolean redelivered)
    throws JMSException;

    public int getJMSPriority() throws JMSException;
    public void setJMSPriority(int priority) throws JMSException;

    public Destination getJMSReplyTo() throws JMSException;
    public void setJMSReplyTo(Destination replyTo) throws JMSException;

    public String getJMSCorrelationID() throws JMSException;
    public void setJMSCorrelationID(String correlationID)
    throws JMSException;

    public byte[] getJMSCorrelationIDAsBytes() throws JMSException;
    public void setJMSCorrelationIDAsBytes(byte[] correlationID)
    throws JMSException;

    public String getJMSType() throws JMSException;
    public void setJMSType(String type) throws JMSException;
}
```

JMS headers are divided into two large groups: automatically assigned headers and developer-assigned headers. The next two sections discuss these two types.

Automatically Assigned Headers

Most JMS headers are automatically assigned; their value is set by the JMS provider when the message is delivered so that values assigned by the developer using the `setJMS<HEADER>()` methods are ignored. In other words, for most headers that are automatically assigned, using the mutator methods explicitly is fruitless.* This doesn't mean, however, that the developer has no control over the value of these headers. Some automatically assigned headers can be programmatically set by the developer when creating the `Session` and `MessageProducer` (i.e., `TopicPublisher`). These cases include

* According to the specification authors, the `setJMS<HEADER>()` methods were left in the `Message` interface for "general orthogonality," or to keep it semantically symmetrical to balance the `getJMS<HEADER>()` methods—a fairly strange but established justification.

the JMSDeliveryMode and the JMSPriority headers, which are illustrated in the header definitions that follow.

JMSDestination

The JMSDestination header identifies the destination with either a Topic or Queue object, both of which are Destination types. Identifying the message's destination is valuable to JMS clients that consume messages from more than one topic or queue:

```
Topic destination = (Topic) message.getJMSDestination();
```

JMSDeliveryMode

There are two types of delivery modes in JMS: persistent and nonpersistent. A persistent message should be delivered *once-and-only-once*, which means that if the JMS provider fails, the message is not lost; it will be delivered after the server recovers. A nonpersistent message is delivered *at-most-once*, which means that it can be lost permanently if the JMS provider fails. In both persistent and nonpersistent delivery modes, the message server should not send a message to the same consumer more than once, but it is possible (see "JMSRedelivered" on page 45 for more details):

```
int deliverymode = message.getJMSDeliveryMode();
if (deliverymode == javax.jms.DeliveryMode.PERSISTENT) {
   ...
} else { // equals DeliveryMode.NON_PERSISTENT
   ...
}
```

The delivery mode can be set using the setJMSDeliveryMode() method on the producer (i.e., TopicPublisher or QueueSender). Once the delivery mode is set on the MessageProducer, it is applied to all messages delivered using that producer. The default setting is PERSISTENT:

```
// Set the JMS delivery mode on the message producer
TopicPublisher topicPublisher = topicSession.createPublisher(topic);
topicPublisher.setDeliveryMode(DeliveryMode.NON_PERSISTENT);
```

JMSMessageID

The JMSMessageID is a String value that uniquely identifies a message. How unique the identifier is depends on the vendor. The JMSMessageID can be useful for historical repositories in JMS consumer applications where messages need to be uniquely indexed. Used in conjunction with the JMSCorrelationID, the JMSMessageID is also useful for correlating messages:

```
String messageid = message.getJMSMessageID();
```

JMSTimestamp

The JMSTimestamp is set automatically by the message producer when the send() operation is invoked. It contains the time the message was received by the JMS provider, not the time it was actually delivered. This header is useful for determining the duration between when the message was sent and when it was actually received by the consumer. The timestamp is a long value that measures time in milliseconds (since January 1, 1970):

```
long timestamp = message.getJMSTimestamp();
```

JMSExpiration

A `Message` object's expiration date prevents the message from being delivered to consumers after it has expired. This is useful for messages whose data is only valid for a period of time:

```
long timeToLive = message.getJMSExpiration();
```

The expiration time for messages is set in milliseconds on the producer (that is, `TopicPublisher`) using the `setTimeToLive()` method:

```
TopicPublisher topicPublisher = topicSession.createPublisher(topic);
// Set time to live as 1 hour (1000 millis x 60 sec x 60 min)
topicPublisher.setTimeToLive(3600000);
```

The provider will then add the `timeToLive` value to the system timestamp and set the `JMSExpiration`. By default the `timeToLive` is zero (0), which indicates that the message doesn't expire. Calling `setTimeToLive()` with a zero argument ensures that a message is created without an expiration date. Any direct programmatic invocation of the `setJMSExpiration()` method will be ignored when the message is sent.

JMSRedelivered

The `JMSRedelivered` header indicates that the message was redelivered to the consumer. The `JMSRedelivered` header is `true` if the message is redelivered, and `false` if it's not. A message may be marked redelivered if a consumer failed to acknowledge previous delivery of the message, or when the JMS provider is not certain whether the consumer has already received the message:

```
boolean isRedelivered = message.getJMSRedelivered()
```

Message redelivery is covered in more detail in Chapter 7.

JMSPriority

The message producer may assign a priority to a message when it is delivered. There are two categories of message priorities: levels 0–4 are gradations of *normal* priority, and levels 5–9 are gradations of *expedited* priority. The message servers may use a message's priority to prioritize delivery of messages to consumers—messages with an expedited priority are delivered ahead of normal priority messages:

```
int priority = message.getJMSPriority();
```

The priority of messages can be declared by the JMS client using the `setPriority()` method on the producer:

```
TopicPublisher topicPublisher = TopicSession.createPublisher(someTopic);
topicPublisher.setPriority(9);
```

Any direct programmatic invocation of the `setJMSPriority()` method will be ignored when the message is sent.

Developer-Assigned Headers

While many of the JMS headers are set automatically when the message is delivered, several others must be set explicitly on the `Message` object before it is delivered by the producer.

JMSReplyTo

In some cases, a JMS message producer may want the consumers to reply to a message. The `JMSReplyTo` header, which contains a `javax.jms.Destination`, indicates which address a JMS consumer should reply to. Using this header property further decouples the message producer from the message consumer when using a request/reply scenario. Note that a JMS consumer is not required to send a reply just because this header property is set:

```
message.setJMSReplyTo(topic);
...
Topic topic = (Topic) message.getJMSReplyTo();
```

JMSCorrelationID

The `JMSCorrelationID` provides a header for associating the current message with some previous message or application-specific ID. In most cases, the `JMSCorrelationID` will be used to tag a message as a reply to a previous message identified by a `JMSMessageID`, but the `JMSCorrelationID` can be any value, not just a `JMSMessageID`:

```
message.setJMSCorrelationID(identifier);
...
String correlationid = message.getJMSCorrelationID();
```

JMSType

`JMSType` is an optional header that is set by the JMS client. Its main purpose is to identify the message structure and type of payload. Note that this header does not indicate the *class* of message being sent (`MapMessage`, `TextMessage`, etc.), but rather an entry in an internal repository used by the JMS provider. Some MOM systems (IBM's WebSphere MQ, for example) treat the message body as uninterpreted bytes. These systems often provide a message type as a simple way for applications to label the message body. So a message type can be useful when exchanging messages with non-JMS clients that require this type of information to process the payload.

Other MOM and EDI systems directly tie each message to some form of external message schema, with the message type as the link. These MOM and EDI systems require a message type because they provide metadata services bound to it. For example, IBM's WebSphere Business Integration (WBI) Adapter Framework uses the `JMSType` to determine what kind of message is being received; a null value indicates an administrative message, whereas values beginning with `mcd://mrm` or `mcd://xml` refer to a business object.

Properties

Properties act like additional headers that can be assigned to a message. They allow the developer to add additional opaque information about the message. They are also used to expose data used for message selectors when doing message filtering. The `Message` interface provides several accessor and mutator methods for reading and writing properties. The value of a property can be a `String`, `boolean`, `byte`, `double`, `int`, `long`, or `float`.

There are three basic categories of message properties: application-specific properties, JMS-defined properties, and provider-specific properties. Application properties are defined and applied to `Message` objects by the application developer; the JMS extension and provider-specific properties are additional headers that are, for the most part, automatically added by the JMS provider.

Application-Specific Properties

Any property defined by the application developer can be an application-specific property. Application properties are set before the message is delivered. There are no predefined application properties; developers are free to define any properties that fit their needs. For example, in the chat example developed in Chapter 2, a special property could be added that identifies the user sending the message:

```
TextMessage message = pubSession.createTextMessage();
message.setText(text);
message.setStringProperty("username",username);
publisher.publish(message);
```

As an application-specific property, `username` is not meaningful outside the `Chat` application; it is used exclusively by the application to filter messages based on the identity of the publisher.

Property values can be any `boolean`, `byte`, `short`, `int`, `long`, `float`, `double`, or `String`. The `javax.jms.Message` interface provides accessor and mutator methods for each of these property value types. Here is a subset of the `Message` interface definition that shows these methods:

```
package javax.jms;

public interface Message {
```

```
    public String getStringProperty(String name)
       throws JMSException, MessageFormatException;
    public void setStringProperty(String name, String value)
       throws JMSException, MessageNotWriteableException;

    public int getIntProperty(String name)
       throws JMSException, MessageFormatException;
    public void setIntProperty(String name, int value)
       throws JMSException, MessageNotWriteableException;

    public boolean getBooleanProperty(String name)
       throws JMSException, MessageFormatException;
    public void setBooleanProperty(String name, boolean value)
       throws JMSException, MessageNotWriteableException;

    public double getDoubleProperty(String name)
       throws JMSException, MessageFormatException;
    public void setDoubleProperty(String name, double value)
       throws JMSException, MessageNotWriteableException;

    public float getFloatProperty(String name)
       throws JMSException, MessageFormatException;
    public void setFloatProperty(String name, float value)
       throws JMSException, MessageNotWriteableException;

    public byte getByteProperty(String name)
       throws JMSException, MessageFormatException;
    public void setByteProperty(String name, byte value)
       throws JMSException, MessageNotWriteableException;

    public long getLongProperty(String name)
       throws JMSException, MessageFormatException;
    public void setLongProperty(String name, long value)
       throws JMSException, MessageNotWriteableException;

    public short getShortProperty(String name)
       throws JMSException, MessageFormatException;
    public void setShortProperty(String name, short value)
       throws JMSException, MessageNotWriteableException;

    public Object getObjectProperty(String name)
       throws JMSException, MessageFormatException;
    public void setObjectProperty(String name, Object value)
       throws JMSException, MessageNotWriteableException;

    public void clearProperties()
       throws JMSException;

    public Enumeration getPropertyNames()
       throws JMSException;

    public boolean propertyExists(String name)
       throws JMSException;
    ...
}
```

The object property methods (`setObjectProperty()` and `getObjectProperty()`) can only be used with object wrappers that correspond to the allowed primitive types (`java.lang.Integer`, `java.lang.Double`, etc.) and the `String` type. They cannot be used with other types of Java objects (such as transfer objects or business objects).

Once a message is published or sent, its properties become read-only; the properties cannot be changed by either the consumer or the producer. If the consumer attempts to set a property, the method throws a `javax.jms.MessageNotWriteableException`. The properties can, however, be changed on that message by calling the method `clearProperties()`, which removes all the properties from the message so that new ones can be added.

The `getPropertyNames()` method in the `Message` interface can be used to obtain an `Enumeration` of all the property names contained in the message. These names can then be used to obtain property values using the property accessor methods. For example:

```
public void onMessage(Message message) {
    Enumeration propertyNames = message.getPropertyNames();
    while(propertyNames.hasMoreElements()) {
       String name = (String)propertyNames.nextElement();
       Object value = getObjectProperty(name);
       System.out.println("\nname+" = "+value);
    }
}
```

JMS-Defined Properties

JMS-defined properties have the same characteristics as application properties, except that most of them are set by the JMS provider when the message is sent. JMS-defined properties act as optional JMS headers; with some noted exceptions, vendors can choose to support none, some, or all of them. The following is a list of the nine JMS-defined properties, which are described in more detail in Appendix C:

- `JMSXUserID`
- `JMSXAppID`
- `JMSXProducerTXID`
- `JMSXConsumerTXID`
- `JMSXRcvTimestamp`
- `JMSXDeliveryCount`
- `JMSXState`
- `JMSXGroupID`
- `JMSXGroupSeq`

Of this list, only `JMSXGroupID` and `JMSXGroupSeq` are required to be supported by all JMS providers. These optional properties are used to group messages.

Note that in the Message interface you will not find corresponding setJMSX<PROPERTY>() and getJMSX<PROPERTY>() methods defined; when used, they must be set in the same manner as application-specified properties:

```
message.setStringProperty("JMSXGroupID", "ERF-001");
message.setIntProperty("JMSXGroupSeq", 3);
```

Provider-Specific Properties

Every JMS provider can define a set of proprietary properties that can be set by the client or the provider automatically. Provider-specific properties must start with the prefix JMS followed by the property name (JMS_<vendor-property-name>). The purpose of the provider-specific properties is to support proprietary vendor features.

Message Types

The Java Message Service defines six Message interface types that must be supported by JMS providers. Although JMS defines the Message interfaces, it doesn't define their implementation. This allows vendors to implement and transport messages in their own way, while maintaining a consistent and standard interface for the JMS application developer. The six message interfaces are Message and its five subinterfaces: TextMessage, StreamMessage, MapMessage, ObjectMessage, and BytesMessage.

The Message interfaces are defined according to the kind of payload they are designed to carry. In some cases, Message types were included in JMS to support legacy payloads that are common and useful, which is the case with the TextMessage, BytesMessage, and StreamMessage message types. In other cases, the Message types were defined to facilitate emerging needs; for example, ObjectMessage can transport serializable Java objects. Some vendors may provide other proprietary message types. Progress's SonicMQ, for example, provides an XMLMessage type that extends the TextMessage, allowing developers to deal with the message directly through DOM or SAX interfaces.

Message

The simplest type of message is the javax.jms.Message, which serves as the base interface to the other message types. As shown below, the Message type can be created and used as a JMS message with no payload:

```
// Create and deliver a Message
Message message =  session.createMessage();
publisher.publish(message);
...

// Receive a message on the consumer
public void onMessage(Message message) {
   // No payload, just process event notification
}
```

This type of message contains only JMS headers and properties and is used in event notification. An event notification is a broadcast, warning, or notice of some occurrence. If the business scenario requires a simple notification without a payload, the lightweight `Message` type is the most efficient way to implement it. For example, to provide a broadcast notification of an exception in a particular class, you could publish a message containing the text of the exception without a payload, as follows:

```
//send the exception
   ...
   try {
      ...
   } catch (Exception up) {
      Message message = session.createMessage();
      message.setStringProperty("Exception", up.getMessage());
      publisher.publish(message);
      throw up;
   }


//receive the exception
   ...
   public void onMessage(Message message) {
      ...
      System.out.println("Exception: " + message.getStringProperty());
      ...
   }
   ...
```

TextMessage

This type carries a `java.lang.String` as its payload. It's useful for exchanging simple text messages and more complex character data like XML documents:

```
package javax.jms;

public interface TextMessage extends Message {
    public String getText()
      throws JMSException;
    public void setText(String payload)
      throws JMSException, MessageNotWriteableException;
}
```

Text messages can be created with one of two factory methods defined in the `Session` interface. One factory method takes no arguments, resulting in a `TextMessage` object with an empty payload—the payload is added using the `setText()` method defined in the `TextMessage` interface. The other factory method takes a `String` type payload as an argument, producing a ready-to-deliver `TextMessage` object:

```
TextMessage textMessage  = session.createTextMessage();
textMessage.setText("Hello!");
topicPublisher.publish(textMessage);
...
```

```
TextMessage textMessage = session.createTextMessage("Hello!");
queueSender.send(textMessage);
```

When a consumer receives a `TextMessage` object, it can extract the `String` payload using the `getText()` method. If the `TextMessage` was delivered without a payload, the `getText()` method returns a `null` value.

ObjectMessage

This type carries a serializable Java object as its payload. It's useful for exchanging Java objects:

```
package javax.jms;

public interface ObjectMessage extends Message {
    public java.io.Serializable getObject()
       throws JMSException;
    public void setObject(java.io.Serializable payload)
       throws JMSException, MessageNotWriteableException;
}
```

Object messages can be created with one of two factory methods defined in the `Session` interface. One factory method takes no arguments, so the serializable object must be added using the `setObject()`. The other factory method takes the `Serializable` payload as an argument, producing a ready-to-deliver `ObjectMessage`:

```
// Order is a serializable object
Order order = new Order( );
...
ObjectMessage objectMessage  = session.createObjectMessage();
objectMessage.setObject(order);
queueSender.send(objectMessage);
...
ObjectMessage objectMessage = session.createObjectMessage(order);
topicPublisher.publish(objectMessage);
```

When a consumer receives an `ObjectMessage`, it can extract the payload using the `getObject()` method. If the `ObjectMessage` was delivered without a payload, the `getObject()` method returns a `null` value:

```
public void onMessage(Message message) {
  try {
    ObjectMessage objectMessage = (ObjectMessage)message;
    Order order = (Order)objectMessage.getObject();
    ...
  } catch (JMSException jmse) {
  ...
}
```

While the ObjectMessage looks very convenient, there are implications to using this message type. For example, when using the ObjectMessage, both the message producer and the message consumer must be JMS-compatible (i.e., written for the Java platform).†

Furthermore, the class definition of the object payload has to be available to both the JMS producer and JMS consumer. If the Order class used in the previous example is not available to the JMS consumer's Java Virtual Machine (JVM), an attempt to access the Order object from the message's payload would result in a java.lang.ClassNotFoundException. Some JMS providers may provide dynamic class-loading capabilities, but that would be a vendor-specific quality of service. Most of the time, the class must be placed on the JMS consumer's class path manually by the developer.

BytesMessage

This type carries an array of primitive bytes as its payload. It's useful for exchanging data in an application's native format, which may not be compatible with other existing Message types. It is also useful where JMS is used purely as a transport between two systems, and the message payload is opaque to the JMS client:

```
package javax.jms;

public interface BytesMessage extends Message {

    public byte readByte() throws JMSException;
    public void writeByte(byte value) throws JMSException;
    public int readUnsignedByte() throws JMSException;

    public int readBytes(byte[] value) throws JMSException;
    public void writeBytes(byte[] value) throws JMSException;
    public int readBytes(byte[] value, int length)
       throws JMSException;
    public void writeBytes(byte[] value, int offset, int length)
       throws JMSException;

    public boolean readBoolean() throws JMSException;
    public void writeBoolean(boolean value) throws JMSException;

    public char readChar() throws JMSException;
    public void writeChar(char value) throws JMSException;

    public short readShort() throws JMSException;
    public void writeShort(short value) throws JMSException;
    public int readUnsignedShort() throws JMSException;

    public void writeInt(int value) throws JMSException;
    public int readInt() throws JMSException;
```

† It's possible that a JMS provider could use the CORBA 2.3 IIOP protocol, which can handle ObjectMessage types consumed by non-Java, non-JMS clients.

```
    public void writeLong(long value) throws JMSException;
    public long readLong() throws JMSException;

    public float readFloat() throws JMSException;
    public void writeFloat(float value) throws JMSException;

    public double readDouble() throws JMSException;
    public void writeDouble(double value) throws JMSException;

    public String readUTF() throws JMSException;
    public void writeUTF(String value) throws JMSException;

    public void writeObject(Object value) throws JMSException;

    public void reset() throws JMSException;
}
```

If you've worked with the `java.io.DataInputStream` and `java.io.DataOutputStream` classes, then the methods of the `BytesMessage` interface, which are loosely based on these I/O classes, will look familiar to you. Most of the methods defined in the `BytesMessage` interface allow the application developer to read and write data to a byte stream using Java's primitive data types. When a Java primitive is written to the `BytesMessage` using one of the `set<TYPE>()` methods, the primitive value is converted to its byte representation and appended to the stream. Here's how a `BytesMessage` is created and how values are written to its byte stream:

```
BytesMessage bytesMessage = session.createBytesMessage();

bytesMessage.writeChar('R');
bytesMessage.writeInt(10);
bytesMessage.writeUTF("OReilly");

queueSender.send(bytesMessage);
```

When a `BytesMessage` is received by a JMS consumer, the payload is a raw byte stream, so it is possible to read the stream using arbitrary types, but this will probably result in erroneous data. It's best to read the `BytesMessage`'s payload in the same order, and with the same types, with which it was written:

```
public void onMessage(Message message) {
  try {
    BytesMessage bytesMessage = (BytesMessage)message;
    char   c = bytesMessage.readChar();
    int    i = bytesMessage.readInt();
    String s = bytesMessage.readUTF();
  } catch (JMSException jmse){
  ...
}
```

In order to read and write `String` values, the `BytesMessage` uses methods based on the UTF-8 format, which is a standard format for transferring and storing Unicode text data efficiently.

> ### UTF-8
>
> UTF-8 encodes Unicode characters as 1 to 4 bytes. The encoding is designed for processing efficiency, ease of implementation in most existing software, and compatibility with ASCII. That is, the encoding of any character in the ASCII character set has the same encoding in UTF-8. So the letter "A" (0×41 in ASCII) is also a 1-byte character with the same value, 0×41, in UTF-8. The characters in ISO 8859-1 that are above the ASCII range (i.e., above 127), when converted to Unicode and encoded in UTF-8, are 2-byte characters.
>
> Every character in the Unicode character set can be expressed in UTF-8, and there is an algorithmic conversion between the 16-bit (2-byte) form of Unicode and UTF-8 that ensures lossless transformations.
>
> One of the key benefits of using the UTF-8 encoding is that null bytes are only used as string terminators. Some mail systems and network protocols cannot tolerate null bytes in the input stream, so the 16-bit encoding of Unicode, which might have a null value in either byte, is unacceptable for these purposes.

The methods for accessing the `short` and `byte` primitives include *unsigned* methods (`readUnsignedShort()`, `readUnsignedByte()`). These methods are something of a surprise, since the `short` and `byte` data types in Java are almost always signed. The values that can be taken by unsigned `byte` and `short` data are what you'd expect: to 255 for a `byte`, and to 65535 for a `short`. Because these values can't be represented by the (signed) `byte` and `short` data types, `readUnsignedByte()` and `readUnsignedShort()` both return an `int`.

In addition to the methods for accessing primitive data types, the `BytesMessage` includes a single `writeObject()` method. This is used for `String` objects and the primitive wrappers: `Byte`, `Boolean`, `Character`, `Short`, `Integer`, `Long`, `Float`, `Double`. When written to the `BytesMessage`, these values are converted to the byte form of their primitive counterparts. The `writeObject()` method is provided as a convenience when the types to be written aren't known until runtime.

If an exception is thrown while reading the `BytesMessage` and you are able to recover from the exception without having to retrieve the message again, the pointer in the stream must be reset to the position it had just been in prior to the read operation that caused the exception. This allows the JMS client to recover from read errors without losing its place in the stream.

The `reset()` method returns the stream pointer to the beginning of the stream and puts the `BytesMessage` in read-only mode so that the contents of its byte stream cannot be further modified. This method can be called explicitly by the JMS client if needed, but it's always called implicitly when the `BytesMessage` is delivered.

The `BytesMessage` is one of the most portable of the message types, and is therefore useful when communicating with non-JMS messaging clients. In some cases, a JMS client may be a kind of router, consuming messages from one source and delivering them to a destination. Routing applications may not need to know the contents of the data they transport and so may choose to transfer payloads as binary data, using a `BytesMessage`, from one location to another.

StreamMessage

The `StreamMessage` carries a stream of primitive Java types (`int`, `double`, `char`, etc.) as its payload. It provides a set of convenience methods for mapping a formatted stream of bytes to Java primitives. Primitive types are read from the `Message` in the same order they were written. Here's the definition of the `StreamMessage` interface:

```
public interface StreamMessage extends Message {

    public boolean readBoolean() throws JMSException;
    public void writeBoolean(boolean value) throws JMSException;

    public byte readByte() throws JMSException;
    public int readBytes(byte[] value) throws JMSException;
    public void  writeByte(byte value) throws JMSException;
    public void writeBytes(byte[] value) throws JMSException;
    public void writeBytes(byte[] value, int offset, int length)
       throws JMSException;

    public short readShort() throws JMSException;
    public void writeShort(short value) throws JMSException;

    public char readChar() throws JMSException;
    public void writeChar(char value) throws JMSException;

    public int readInt() throws JMSException;
    public void writeInt(int value) throws JMSException;

    public long readLong() throws JMSException;
    public void writeLong(long value) throws JMSException;

    public float readFloat() throws JMSException;
    public void writeFloat(float value) throws JMSException;

    public double readDouble() throws JMSException;
    public void  writeDouble(double value) throws JMSException;

    public String readString() throws JMSException;
    public void writeString(String value) throws JMSException;

    public Object readObject() throws JMSException;
    public void writeObject(Object value) throws JMSException;

    public void reset() throws JMSException;
}
```

On the surface, the `StreamMessage` strongly resembles the `BytesMessage`, but they are not the same. The `StreamMessage` keeps track of the order and types of primitives written to the stream, so formal conversion rules apply. For example, an exception would be thrown if you tried to read a `long` value as a `short`:

```
StreamMessage streamMessage = session.createStreamMessage();
streamMessage.writeLong(2938302);

// The next line throws a JMSException
short value = streamMessage.readShort();
```

While this would work fine with a `BytesMessage`, it won't work with a `StreamMessage`. A `BytesMessage` would write the `long` as 64 bits (8 bytes) of raw data, so that you could later read some of the data as a `short`, which is only 16 bits (the first 2 bytes of the long). The `StreamMessage`, on the other hand, writes the type information as well as the value of the `long` primitive, and enforces a strict set of conversion rules that prevent reading the `long` as a `short`.

Table 3-1 shows the conversion rules for each type. The left column shows the type written, and the right column shows how that type may be read. A `JMSException` is thrown by the accessor methods to indicate that the original type could not be converted to the type requested. This is the exception that would be thrown if you attempted to read `long` as a `short`.

Table 3-1. Type conversion rules

write<TYPE>()	read<TYPE>()
boolean	boolean, String
byte	byte, short, int, long, String
short	short, int, long, String
char	char, String
long	long, String
int	int, long, String
float	float, double, String
double	double, String
String	String, boolean, byte, short, int, long, float, double
byte []	byte []

`String` values can be converted to any primitive data type if they are formatted correctly. If the `String` value cannot be converted to the primitive type requested, a `java.lang.NumberFormatException` is thrown. However, most primitive values can be accessed as a `String` using the `readString()` method. The only exceptions to this rule are `char` values and `byte` arrays, which cannot be read as `String` values.

The `writeObject()` method follows the rules outlined for the similar method in the `BytesMessage` class. Primitive wrappers are converted to their primitive counterparts. The `readObject()` method returns the appropriate object wrapper for primitive values, or a String or a byte array, depending on the type that was written to the stream. For example, if a value was written as a primitive int, it can be read as a java.lang.Integer object.

The StreamMessage also allows null values to be written to the stream. If a JMS client attempts to read a null value using the readObject() method, null is returned. The rest of the primitive accessor methods attempt to convert the null value to the requested type using the valueOf() operations. The readBoolean() method returns false for null values, while the other primitive property methods throw the java.lang.NumberFormatException. The readString() method returns null or possibly an empty String ("") depending on the implementation. The readChar() method throws a NullPointerException.

If an exception is thrown while reading the StreamMessage and you are able to recover from the exception without having to retrieve the message again, the pointer in the stream is reset to the position it had just been in prior to the read operation that caused the exception. This allows the JMS client to recover gracefully from exceptions without losing the pointer's position in the stream.

The reset() method returns the stream pointer to the beginning of the stream and puts the message in a read-only mode. It is called automatically when the message is delivered to the client. However, it may need to be called directly by the consuming client when a message is redelivered:

```
if ( strmMsg.getJMSRedelivered() )
    strmMsg.reset();
```

MapMessage

This type carries a set of *name-value* pairs as its payload. The payload is similar to a java.util.Properties object, except the values must be Java primitives (or their wrappers) in addition to Strings. The MapMessage class is useful for delivering keyed data that may change from one message to the next:

```
public interface MapMessage extends Message {

    public boolean getBoolean(String name) throws JMSException;
    public void setBoolean(String name, boolean value)
       throws JMSException;

    public byte getByte(String name) throws JMSException;
    public void setByte(String name, byte value) throws JMSException;

    public byte[] getBytes(String name) throws JMSException;
    public void setBytes(String name, byte[] value)
       throws JMSException;
    public void setBytes(String name, byte[] value, int offset, int length)
       throws JMSException;
```

```
    public short getShort(String name) throws JMSException;
    public void setShort(String name, short value) throws JMSException;

    public char getChar(String name) throws JMSException;
    public void setChar(String name, char value) throws JMSException;

    public int getInt(String name) throws JMSException;
    public void setInt(String name, int value) throws JMSException;

    public long getLong(String name) throws JMSException;
    public void setLong(String name, long value) throws JMSException;

    public float getFloat(String name) throws JMSException;
    public void setFloat(String name, float value)
       throws JMSException;

    public double getDouble(String name) throws JMSException;
    public void setDouble(String name, double value)
       throws JMSException;

    public String getString(String name) throws JMSException;
    public void setString(String name, String value)
       throws JMSException;

    public Object getObject(String name) throws JMSException;
    public void setObject(String name, Object value)
       throws JMSException;

    public Enumeration getMapNames() throws JMSException;
    public boolean itemExists(String name) throws JMSException;
}
```

Essentially, MapMessage works similarly to JMS properties: any name-value pair can be written to the payload. The name must be a String object, and the value may be a String or a primitive type. The values written to the MapMessage can then be read by a JMS consumer using the name as a key:

```
MapMessage mapMessage = session.createMapMessage();
mapMessage.setInt("Age", 88);
mapMessage.setFloat("Weight", 234);
mapMessage.setString("Name", "Smith");
mapMessage.setObject("Height", new Double(150.32));
....

int age = mapMessage.getInt("Age");
float weight = mapMessage.getFloat("Weight");
String name = mapMessage.getString("Name");
Double height = (Double)mapMessage.getObject("Height");
```

The setObject() method writes a Java primitive wrapper type, String object, or byte array. The primitive wrappers are converted to their corresponding primitive types when set. The getObject() method reads Strings, byte arrays, or any primitive type as its corresponding primitive wrapper.

The conversion rules defined for the StreamMessage apply to the MapMessage. See Table 3-1 in the section "StreamMessage" on page 56.

A JMSException is thrown by the accessor methods to indicate that the original type could not be converted to the type requested. In addition, String values can be converted to any primitive value type if they are formatted correctly; the accessor will throw a java.lang.NumberFormatException if they aren't.

If a JMS client attempts to read a name-value pair that doesn't exist, the value is treated as if it was null. Although the getObject() method returns null for nonexistent mappings, the other types behave differently. While most primitive accessors throw the java.lang.NumberFormatException if a null value or nonexistent mapping is read, other accessors behave as follows: the getBoolean() method returns false for null values; the getString() returns a null value or possibly an empty String (""), depending on the implementation; and the getChar() method throws a NullPointerException.

To avoid reading nonexistent name-value pairs, the MapMessage provides an itemExists() test method. In addition, the getMapNames() method lets a JMS client enumerate the names and use them to obtain all the values in the message. For example:

```
public void onMessage(Message message) {
   MapMessage mapMessage = (MapMessage)message;
   Enumeration names = mapMessage.getMapNames();
   while(names.hasMoreElements()){
      String name = (String)names.nextElement();
      Object value = mapMessage.getObject(name);
      System.out.println("Name = "+name+", Value = "+value);
   }
}
```

Read-Only Messages

When messages are delivered, the body of the message is made read-only. Any attempt to alter a message body after it has been delivered results in a javax.jms.MessageNotWriteableException. The only way to change the body of a message after it has been delivered is to invoke the clearBody() method, which is defined in the Message interface. The clearBody() method empties the message's payload so that a new payload can be added.

Properties are also read-only after a message is delivered. Why are both the body and properties made read-only after delivery? It allows the JMS provider more flexibility in implementing the Message object. For example, a JMS provider may choose to stream a BytesMessage or StreamMessage as it is read, rather than all at once. Another vendor may choose to keep properties or body data in an internal buffer so that it can be read directly without the need to make a copy, which is especially useful with multiple consumers on the same client.

Client-Acknowledged Messages

The `acknowledge()` method, defined in the `Message` interface, is used when the consumer has chosen `CLIENT_ACKNOWLEDGE` as its acknowledgment mode. There are three acknowledgment modes that may be set by the JMS consumer when its session is created: `AUTO_ACKNOWLEDGE`, `DUPS_OK_ACKNOWLEDGE`, and `CLIENT_ACKNOWLEDGE`. Here is how a pub/sub consumer sets one of the three acknowledgment modes:

```
TopicSession topic =
   topicConnection.createTopicSession(false, Session.CLIENT_ACKNOWLEDGE);
```

In `CLIENT_ACKNOWLEDGE` mode, the JMS consumer (client) explicitly acknowledges each message as it is received. The `acknowledge()` method on the `Message` interface is used for this purpose. For example:

```
public void onMessage(Message message) {
   message.acknowledge();
   ...
}
```

The other acknowledgment modes do not require the use of this method and are covered in more detail in Chapter 7 and Appendix B.

Any acknowledgment mode specified for a transacted session is ignored. When a session is transacted, the acknowledgment is part of the transaction and is executed automatically prior to the commit of the transaction. If the transaction is rolled back, no acknowledgment is given. Transactions are covered in more detail in Chapter 7.

Interoperability and Portability of Messages

A message delivered by a JMS client may be converted to a JMS provider's native format and delivered to non-JMS clients, but it must still be consumable as its original `Message` type by JMS clients. Messages delivered from non-JMS clients to a JMS provider may be consumable by JMS clients—the JMS provider should attempt to map the message to its closest JMS type or, if that's not possible, to the `BytesMessage`.

JMS providers are not required to be interoperable. A message published to one JMS provider's server is not consumable by another JMS provider's consumer. In addition, a JMS provider usually can't publish or read messages from destinations (topics and queues) implemented by another JMS provider. Most JMS providers have bridges or connectors to address this issue.

Although interoperability is not required, limited message portability is required. A message consumed or created using JMS provider P1 can be delivered using JMS provider P2. JMS provider P2 will simply use the accessor methods of the message to read its headers, properties, and payload and convert them to its own native format: not a fast process, but portable. This portability is limited to interactions of the JMS client, which takes a message from one provider and passes it to another.

CHAPTER 4

Point-to-Point Messaging

This chapter focuses on the point-to-point (p2p) messaging model. The point-to-point model is used when you need to send a message to only one message consumer. Even though multiple consumers may be listening on the queue for the same message, only one of those consumer threads will receive the message. This is different from the publish-and-subscribe model described in Chapter 5, where a message is broadcast to (and consumed by) multiple consumers.

In this chapter, we will describe the point-to-point model through the use of a typical messaging scenario involving a borrower and a mortgage lender. In our example, the `QBorrower` class will submit a loan application via JMS messaging to a `QLender` class. The `QLender` class will receive the loan request through a message queue, determine whether to accept or decline the loan based on certain business rules, and send the result (accept or decline) back to the `QBorrower` class through another message queue. However, before launching into the messaging example, we will first describe the main characteristics and use cases of the p2p messaging model.

Point-to-Point Overview

In the p2p model, the producer is called a *sender* and the consumer is called a *receiver*. The most important characteristics of the point-to-point model are as follows:

- Messages are exchanged through a virtual channel called a *queue*. A queue is a destination to which producers send messages and a source from which receivers consume messages.
- Each message is delivered to only one receiver. Multiple receivers may listen on a queue, but each message in the queue may only be consumed by one of the queue's receivers.
- Messages are ordered. A queue delivers messages to consumers in the order they were placed in the queue by the message server. As messages are consumed, they are removed from the head of the queue (unless message priority is used).

- There is no coupling of the producers to the consumers. Receivers and senders can be added dynamically at runtime, allowing the system to grow or shrink in complexity over time. (This is a characteristic of messaging systems in general.)

Point-to-point messaging is based on the concept of sending a message to a named destination. The actual network location of the destination is transparent to the sender, because the p2p client works with a `Queue` identifier obtained from a JNDI namespace.

As you will see in the next chapter, the pub/sub model is based on a push model, which means that consumers are delivered messages without having to request them. With the p2p messaging model, the p2p receiver can either push or pull messages, depending on whether it uses the asynchronous `onMessage()` callback or a synchronous `receive()` method. Both of these methods are explained in more detail later in this chapter.

In the p2p model there is no direct coupling of the producers to the consumers. The destination queue provides a virtual channel that decouples consumers from producers. In the pub/sub model, multiple consumers that subscribe to the same topic each receive their own copy of every message addressed to that topic. In the p2p model, multiple consumers can use the same queue, but each message delivered to the queue can only be received by one of the queue's consumers. How messages sent to a queue are distributed to the queue's consumers depends on the policies of the JMS provider. Some JMS providers use load-balancing techniques to distribute messages evenly among consumers, while others will use more arbitrary policies.

Messages intended for a p2p queue can be either persistent or nonpersistent. Persistent messages survive JMS provider failures, while nonpersistent messages do not. Messages may also have a priority and an expiration time. One important difference between point-to-point and publish/subscribe messaging is that p2p messages are always delivered, regardless of the current connection status of the receiver. Once a message is delivered to a queue, it stays there even if no consumer is currently connected.

There are two types of point-to-point messaging: asynchronous fire-and-forget processing and asynchronous request/reply processing. With fire-and-forget processing, the message producer sends a message to a queue and does not expect to receive a response (at least not right away). This type of processing can be used to trigger an event or make a request to a receiver to execute a particular action that does not require a response (or in some cases, an immediate response). For instance, you may want to use asynchronous fire-and-forget processing to send a message to a logging system, make a request to kick off a report, or trigger an event on another process. Asynchronous fire-and-forget processing is illustrated in Figure 4-1.

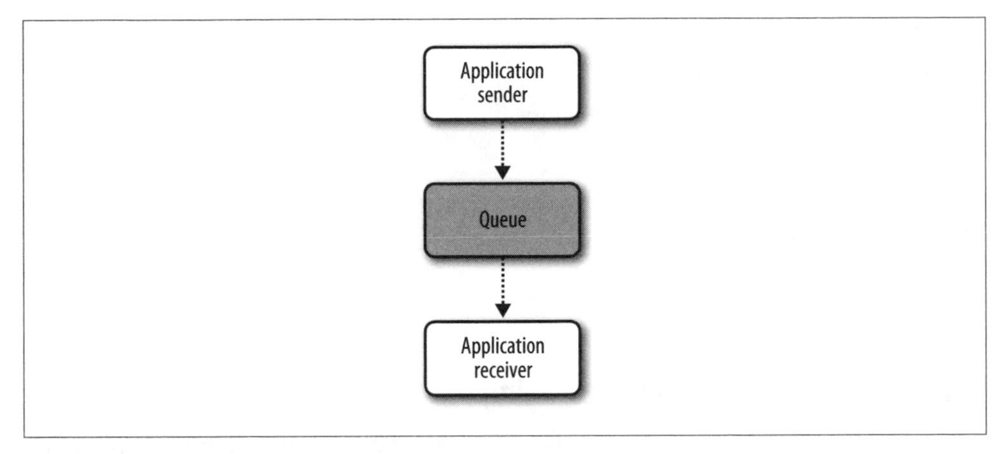

Figure 4-1. p2p async fire-and-forget

With asynchronous request/reply processing, the message producer sends a message on one queue and then does a blocking wait on a reply queue waiting for the response from the receiver. The request/reply processing provides for a high degree of decoupling between the producer and consumer, allowing the message producer and consumer components to be heterogeneous languages or platforms. Asynchronous request/reply processing is illustrated in Figure 4-2.

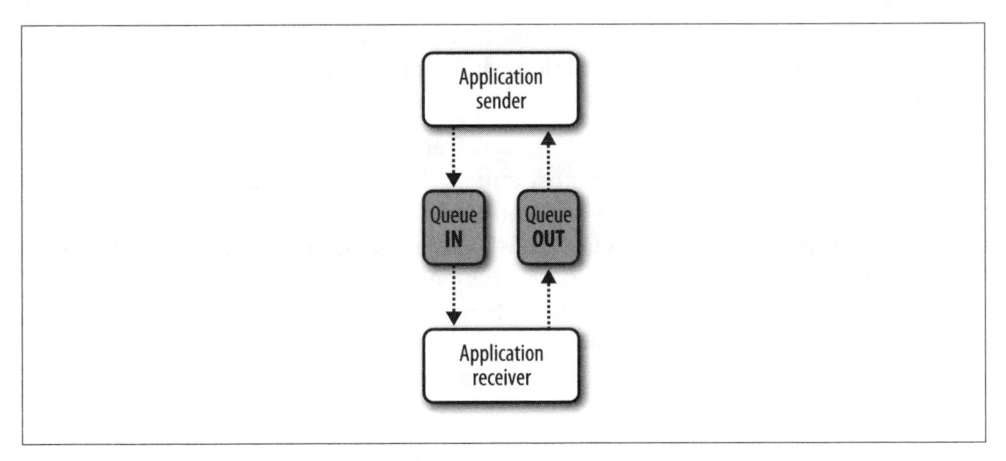

Figure 4-2. p2p async request/reply

The specific p2p interfaces for connecting, creating, sending, and receiving are shown in Table 4-1.

Table 4-1. Interfaces for queues

General API	Point-to-point API
ConnectionFactory	QueueConnectionFactory
Destination	Queue
Connection	QueueConnection
Session	QueueSession
MessageConsumer	QueueSender
MessageProducer	QueueReceiver

When to Use Point-to-Point Messaging

The rationale behind the two models (point-to-point and publish-and-subscribe) lies in the origin of the JMS specification. JMS started out as a way of providing a common API for accessing existing messaging systems. At the time of its conception, some messaging vendors had a p2p model, and some had a pub/sub model. Hence JMS needed to provide an API for both models to gain wide industry support.

In most cases, the decision about which model to use depends on the distinct characteristics of each model. With pub/sub, any number of subscribers can be listening on a topic, all receiving copies of the same message. The publisher generally does not care how many subscribers there are or how many of them are actively listening on the topic. For example, consider a publisher that broadcasts stock quotes. If any particular subscriber is not currently connected and misses out on a great quote, the publisher is not concerned. In contrast, with point-to-point messaging, a particular message is likely to be intended for a one-on-one conversation with a specific application at the other end. In this scenario, every message matters.

Point-to-point is used when you want one receiver to process any given message once and only once. This is perhaps the most critical difference between the two models: point-to-point guarantees that only one consumer will process a given message. This is extremely important when messages need to be processed separately but in tandem, balancing the load of message processing across many JMS clients. Another advantage is that the point-to-point model provides a `QueueBrowser` that allows the JMS client to take a snapshot of the queue to see messages waiting to be consumed. Pub/sub does not include a browsing feature. We'll talk more about the `QueueBrowser` later in this chapter.

Another use case for point-to-point messaging is when you need synchronous communication between components, but those components are written in different programming languages or implemented in different technology platforms (e.g., J2EE and .NET). For example, you may have a stock trading client written as a Java Swing client that needs to communicate with a .NET/C# trading server to process the stock trade. In this scenario, point-to-point messaging can be used to provide the interoperability between these heterogeneous platforms.

As you will see later in this chapter, another good reason to use point-to-point messaging is to provide a higher degree of throughput to server-side components through the use of message-based load balancing, particularly for homogeneous components (i.e., Java to Java). Introducing p2p messaging allows you to add a degree of concurrent processing to your architecture without having to deal with threads or Java concurrency issues.

The QBorrower and QLender Application

To illustrate how point-to-point messaging works, we will use a simple decoupled request/reply example where a `QBorrower` class makes a simple mortgage loan request to a `QLender` class using point-to-point messaging. The `QBorrower` class sends the loan request to the `QLender` class using a `LoanRequest` queue, and based on certain business rules, the `QLender` class sends a response back to the `QBorrower` class using a `LoanResponseQ` queue indicating whether the loan request was approved or denied. Since the `QBorrower` is interested in finding out right away whether the loan was approved or not, once the loan request is sent, the `QBorrower` class will block and wait for a response from the `QLender` class before proceeding. This simple example models a typical messaging request/reply scenario.

Configuring and Running the Application

Before looking at the code, let's look at how the application works. As with the `Chat` application, the `QBorrower` class and `QLender` class both include a `main()` method so they can be run as a standalone Java application. To keep the code vendor-agnostic, both classes need the connection factory name and queue names when starting the application. The `QLender` class is executed from the command line as follows:

```
java ch04.p2p.QLender ConnectionFactory RequestQueue
```

where `ConnectionFactory` is the name of the queue connection factory defined in your JMS provider and `RequestQueue` is the name of the queue that the `QLender` class should be listening on to receive loan requests. As you'll see later in this chapter, the `QBorrower` sends the destination for the response message in the `JMSReplyTo` header property, which is why you do not need to specify it when starting the `QLender` class.

The `QBorrower` class can be executed in the same manner in a separate command window:

```
java ch04.p2p.QBorrower ConnectionFactory RequestQueue ReplyQueue
```

where `ConnectionFactory` is the name of the queue connection factory defined in your JMS provider, `RequestQueue` is the name of the queue that the `QBorrower` class should send loan requests to, and `ReplyQueue` is the name of the queue that the `QBorrower` class should use to receive the results from the `QLender` class.

You will also need to define a *jndi.properties* file in your classpath that contains the JNDI connection information for the JMS provider. The *jndi.properties* file contains the initial context factory class, provider URL, username, and password needed to connect to the JMS server. Each vendor will have a different context factory class and URL name for connecting to the server. You will need to consult the documentation of your specific JMS provider or Java EE container to obtain these values. We have included the steps for configuring ActiveMQ to run the examples in this chapter in Appendix D.

The `QBorrower` and `QLender` classes both require the queue connection factory name and queue names to run. We have chosen to name the connection factory `QueueCF`, and the loan request and loan response queues `LoanRequestQ` and `LoanResponseQ`, respectively. These JNDI resources are typically configured in the JMS provider XML configuration files or configuration screens. You will need to consult your JMS provider documentation on how to configure these resources (please refer to Appendix D for the specific configuration settings for ActiveMQ used to run the examples in this chapter).

You can run the `QBorrower` and `QLender` classes by entering the following two commands in separate command windows:

```
java ch04.p2p.QLender QueueCF LoanRequestQ

java ch04.p2p.QBorrower QueueCF LoanRequestQ LoanResponseQ
```

When the `QBorrower` class starts, you will be prompted to enter a salary amount and the requested loan amount. When you press enter, the `QBorrower` class will send the salary and loan amount to the `QLender` class via the `LoanRequestQ` queue, wait for the response on the `LoanResponseQ` queue, and display whether the loan was approved or denied:

```
QBorrower Application Started
Press enter to quit application
Enter: Salary, Loan_Amount
e.g. 50000, 120000

> 80000, 200000
Loan request was Accepted!

> 50000, 300000
Loan request was Declined
>
```

Here's what happened. The `QBorrower` sent the salary ($80,000) and the loan amount ($200,000) to the `LoanRequestQ` queue, then blocked and waited for a response from the `QLender` class. The `QLender` class received the request on the `LoanRequestQ` queue, applied the simple business logic based on the salary to loan ratio, and sent back the response on the `LoanResponseQ` queue. The message was then received by the `QBorrower` class, and the contents of the return message displayed on the console. This interaction is illustrated in Figure 4-3.

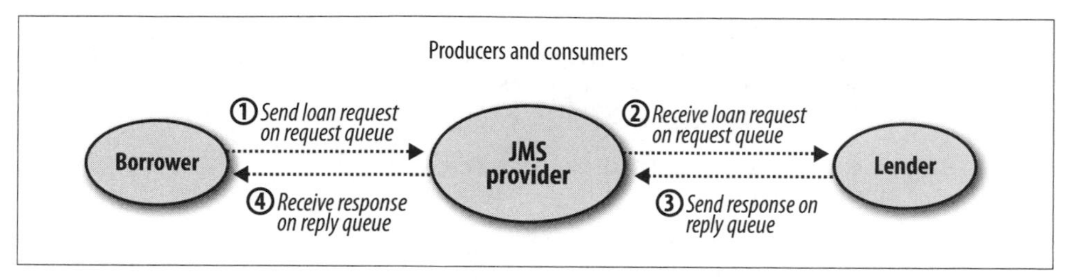

Figure 4-3. Producers and consumers in the loan example

The rest of this chapter examines the source code for the `QBorrower` and `QLender` classes, and covers several advanced subjects related to the point-to-point messaging model.

The QBorrower Class

The `QBorrower` class is responsible for sending a loan request message to a queue containing a salary and loan amount. The class is fairly straightforward: the constructor establishes a connection to the JMS provider, creates a `QueueSession`, and gets the request and response queues using a JNDI lookup. The `main` method instantiates the `QBorrower` class and, upon receiving a salary and loan amount from standard input, invokes the `sendLoanRequest` method to send the message to the queue. Here is the listing for the `QBorrower` class in its entirety. We will be examining the JMS aspects of this class in detail after the full listing:

```
package ch04.p2p;

import java.io.*;
import java.util.StringTokenizer;
import javax.jms.*;
import javax.naming.*;

public class QBorrower {

   private QueueConnection qConnect = null;
   private QueueSession qSession = null;
   private Queue responseQ = null;
   private Queue requestQ = null;

   public QBorrower(String queuecf, String requestQueue,
                    String responseQueue) {
      try {
         // Connect to the provider and get the JMS connection
         Context ctx = new InitialContext();
         QueueConnectionFactory qFactory = (QueueConnectionFactory)
            ctx.lookup(queuecf);
         qConnect = qFactory.createQueueConnection();

         // Create the JMS Session
         qSession = qConnect.createQueueSession(
            false, Session.AUTO_ACKNOWLEDGE);
```

```
         // Lookup the request and response queues
         requestQ = (Queue)ctx.lookup(requestQueue);
         responseQ = (Queue)ctx.lookup(responseQueue);

         // Now that setup is complete, start the Connection
         qConnect.start();

      } catch (JMSException jmse) {
         jmse.printStackTrace();
         System.exit(1);
      } catch (NamingException jne) {
         jne.printStackTrace();
         System.exit(1);
      }
   }

   private void sendLoanRequest(double salary, double loanAmt) {
      try {
         // Create JMS message
         MapMessage msg = qSession.createMapMessage();
         msg.setDouble("Salary", salary);
         msg.setDouble("LoanAmount", loanAmt);
         msg.setJMSReplyTo(responseQ);

         // Create the sender and send the message
         QueueSender qSender = qSession.createSender(requestQ);
         qSender.send(msg);

         // Wait to see if the loan request was accepted or declined
         String filter =
            "JMSCorrelationID = '" + msg.getJMSMessageID() + "'";
         QueueReceiver qReceiver = qSession.createReceiver(responseQ, filter);
         TextMessage tmsg = (TextMessage)qReceiver.receive(30000);
         if (tmsg == null) {
            System.out.println("QLender not responding");
         } else {
            System.out.println("Loan request was " + tmsg.getText());
         }

      } catch (JMSException jmse) {
         jmse.printStackTrace();
         System.exit(1);
      }
   }

   private void exit() {
      try {
         qConnect.close();
      } catch (JMSException jmse) {
         jmse.printStackTrace();
      }
      System.exit(0);
   }
```

```
    public static void main(String argv[]) {
       String queuecf = null;
       String requestq = null;
       String responseq = null;
       if (argv.length == 3) {
          queuecf = argv[0];
          requestq = argv[1];
          responseq = argv[2];
       } else {
          System.out.println("Invalid arguments. Should be: ");
          System.out.println
             ("java QBorrower factory requestQueue responseQueue");
          System.exit(0);
       }

       QBorrower borrower = new QBorrower(queuecf, requestq, responseq);

       try {
          // Read all standard input and send it as a message
          BufferedReader stdin = new BufferedReader
             (new InputStreamReader(System.in));
          System.out.println ("QBorrower Application Started");
          System.out.println ("Press enter to quit application");
          System.out.println ("Enter: Salary, Loan_Amount");
          System.out.println("\ne.g. 50000, 120000");

          while (true) {
             System.out.print("> ");
             String loanRequest = stdin.readLine();
             if (loanRequest == null ||
                 loanRequest.trim().length() <= 0) {
                borrower.exit();
             }

             // Parse the deal description
             StringTokenizer st = new StringTokenizer(loanRequest, ",") ;
             double salary =
                Double.valueOf(st.nextToken().trim()).doubleValue();
             double loanAmt =
                Double.valueOf(st.nextToken().trim()).doubleValue();

             borrower.sendLoanRequest(salary, loanAmt);
          }
       } catch (IOException ioe) {
          ioe.printStackTrace();
       }
    }
}
```

The `main` method of the `QBorrower` class accepts three arguments from the command line: the JNDI name of the queue connection factory, the JNDI name of the loan request queue, and finally, the JNDI name of the loan response queue where the response from the `QLender` class will be received. Once the input parameters have been validated, the

QBorrower class is instantiated and a loop is started that reads the salary and loan amount into the class from the console:

```
String loanRequest = stdin.readLine();
```

The salary and loan amount input data is then parsed, and finally the sendLoanRequest method invoked. The input loop continues until the user presses enter on the console without entering any data:

```
if (loanRequest == null ||
    loanRequest.trim().length() <= 0) {
   borrower.exit();
}
```

Now let's look at the JMS portion of the code in detail, starting with the constructor and ending with the sendLoanRequest method.

JMS Initialization

In the QBorrower class example, all of the JMS initialization logic is handled in the constructor. The first thing the constructor does is establish a connection to the JMS provider by creating an InitialContext:

```
Context ctx = new InitialContext();
```

The connection information needed to connect to the JMS provider is specified in the *jndi.properties* file located in the classpath (see Appendix D for an example). Once we have a JNDI context, we can get the QueueConnectionFactory using the JNDI connection factory name passed into the constructor arguments. The QueueConnectionFactory is then used to create the QueueConnection using a factory method on the QueueConnectionFactory:

```
QueueConnectionFactory qFactory =
   (QueueConnectionFactory) ctx.lookup(queuecf);
qConnect = qFactory.createQueueConnection();
```

Alternatively, you can pass a username and password into the createQueueConnection method as String arguments to perform basic authentication on the connection. A JMSSecurityException will be thrown if the user fails to authenticate:

```
qConnect = qFactory.createQueueConnection("system", "manager");
```

At this point a connection is created to the JMS provider. When the QueueConnection is created, the connection is initially in *stopped mode*. This means you can send messages to the queue, but no message consumers (including the QBorrower class, which is also a message consumer) may receive messages from this connection until it is started.

The QueueConnection object is used to create a JMS Session object (specifically, a QueueSession), which is the working thread and transactional unit of work in JMS. Unlike JDBC, which requires a connection for each transactional unit of work, JMS uses a single connection and multiple Session objects. Typically, applications will create a single JMS Connection on application startup and maintain a pool of Session objects for use whenever a message needs to be produced or consumed.

The QueueSession object is created through a factory object on the QueueConnection object. The QueueConnection variable is declared outside of the constructor in our example so that the connection can be closed in the exit method of the QBorrower class. It is important to close the connection after it is no longer being used to free up resources. Closing the Connection object also closes any open Session objects associated with the connection. The statement in the constructor to create the QueueSession is as follows:

```
qSession =
   qConnect.createQueueSession(false, Session.AUTO_ACKNOWLEDGE);
```

Notice that the createQueueSession method takes two parameters. The first parameter indicates whether the QueueSession is transacted or not. A value of true indicates that the session is transacted, meaning that messages sent to queues during the lifespan of the QueueSession will not be delivered to the receivers until the commit method is invoked on the QueueSession. Likewise, invoking the rollback method on the QueueSession will remove any messages sent during the transacted session. The second parameter indicates the acknowledgment mode. The three possible values are Session.AUTO_ACKNOWLEDGE, Session.CLIENT_ACKNOWLEDGE, and Session.DUPS_OK_ACKNOWLEDGE. The acknowledgment mode is ignored if the session is transacted. Acknowledgment modes are discussed in more detail in Chapter 7.

The next two lines in the constructor perform a JNDI lookup to the JMS provider to obtain the administered destinations. In our case, the JMS destination is cast to a Queue. The argument supplied to each of the lookup methods is a String value containing the JNDI name of the queues we are using in the class:

```
requestQ = (Queue)ctx.lookup(requestQueue);
responseQ = (Queue)ctx.lookup(responseQueue);
```

The final line of code starts the connection, allowing messages to be received on this connection. It is generally a good idea to perform all of your initialization logic before starting the connection:

```
qConnect.start();
```

Interestingly enough, you do not need to start the connection if all you are doing is sending messages. However, it is generally advisable to start the connection to avoid future issues if there is a chance the connection may be shared or request/reply processing added to the sender class.

Another useful thing you can obtain from the JMS Connection is the metadata about the connection. Invoking the getMetaData method on the Connection object gives you a ConnectionMetaData object that provides useful information such as the JMS version, JMS provider name, JMS provider version, and the JMSX property name extensions supported by the JMS provider:

```
import java.util.Enumeration;
import javax.jms.ConnectionMetaData;
import javax.jms.JMSException;
import javax.jms.QueueConnection;
import javax.jms.QueueConnectionFactory;
import javax.naming.Context;
import javax.naming.InitialContext;
import javax.naming.NamingException;

public class MetaData {
  public static void main(String[] args) {
    try {
      Context ctx = new InitialContext();
      QueueConnectionFactory qFactory = (QueueConnectionFactory)
         ctx.lookup("QueueCF");
      QueueConnection qConnect = qFactory.createQueueConnection();
      ConnectionMetaData metadata = qConnect.getMetaData();
      System.out.println("JMS Version:  " +
                         metadata.getJMSMajorVersion() + "." +
                         metadata.getJMSMinorVersion());
      System.out.println("JMS Provider: " +
                         metadata.getJMSProviderName());
      System.out.println("JMSX Properties Supported: ");
      Enumeration e = metadata.getJMSXPropertyNames();
      while (e.hasMoreElements()) {
        System.out.println("   " + e.nextElement());
      }
    } catch (Exception ex) {
      ex.printStackTrace();
      System.exit(1);
    }
  }
}
```

For example, invoking the previous code using the ActiveMQ open source JMS provider will yield the following results:

```
JMS Version:  1.1
JMS Provider: ActiveMQ
JMSX Properties Supported:
   JMSXGroupID
   JMSXGroupSeq
   JMSXDeliveryCount
   JMSXProducerTXID
```

This information can be logged on application startup, indicating the JMS provider and version numbers. It is particularly useful for products or applications that may use multiple providers.

Sending the message and receiving the response

Once the `QBorrower` class is initialized, the salary and loan amounts are entered through the command line. At this point, the `sendLoanRequest` method is invoked from the `main` method to send the loan request to the queue and wait for the response from the

QLender class. At the start of this method, we chose to create a MapMessage but we could have used any of the five JMS message types:

```
MapMessage msg = qSession.createMapMessage();
msg.setDouble("Salary", salary);
msg.setDouble("LoanAmount", loanAmt);
msg.setJMSReplyTo(responseQ);
```

Notice that the JMS message is created from the Session object via a factory method matching the message type. Instantiating a new JMS message object using the new keyword will not work; it must be created from the Session object. After creating and loading the message object, we are also setting the JMSReplyTo message header property to the response queue, which further decouples the producer from the consumer. The practice of setting the JMSReplyTo header property in the message producer as opposed to specifying the reply-to queue in the message consumer is a standard practice when using the request/reply model.

After the message is created, we then create the QueueSender object, specifying the queue we wish to send messages to, and then send the message using the send method:

```
QueueSender qSender = qSession.createSender(requestQ);
qSender.send(msg);
```

There are several overridden send methods available in the QueueSender object. The one we are using here accepts only the JMS message object as the single argument. The other overridden methods allow you to specify the Queue, the delivery mode, the message priority, and finally the message expiry. Since we are not specifying any of the other values in the example just shown, the message priority is set to normal (4), the delivery mode is set to persistent messages (DeliveryMode.PERSISTENT), and the message expiry (time to live) is set to 0, indicating that the message will never expire. All of these parameters can be overridden by using one of the other send methods.

Once the message has been sent, the QBorrower class will block and wait for a response from the QLender on whether the loan was approved or denied. The first step in this process is to set up a message selector so that we can correlate the response message with the one we sent. This is necessary because there may be many other loan requests being sent to and from the loan request queues while we are making our loan request. To make sure we get the proper response back, we would use a technique called *message correlation*. Message correlation is required when using the request/reply model of point-to-point messaging where the queue is being shared by multiple producers and consumers (see "Message Correlation" on page 81 for more details):

```
String filter = "JMSCorrelationID = '" + msg.getJMSMessageID() + "'";
QueueReceiver qReceiver = qSession.createReceiver(responseQ, filter);
```

Notice we specify the filter when creating the QueueReceiver, indicating that we only want to receive messages when the JMSCorrelationID is equal to the original JMSMessageID. Now that we have a QueueReceiver, we can invoke the receive method to do a blocking wait until the response message is received. In this case, we are using the overridden receive method that accepts a timeout value in milliseconds:

```
TextMessage tmsg = (TextMessage)qReceiver.receive(30000);
if (tmsg == null) {
   System.out.println("QLender not responding");
} else {
   System.out.println("Loan request was " + tmsg.getText());
}
```

It is a good idea to always specify a reasonable timeout value on the `receive` method; otherwise, it will sit there and wait forever (in effect, the application would "hang"). Specifying a reasonable timeout value allows the request/reply sender (in this case the `QBorrower`) to take action in the event the message has not been delivered in a timely fashion or there is a problem on the receiving side (in this case the `QLender`). If a timeout condition does occur, the message returned from the `receive` method will be `null`. Note that it is the entire message object that is `null`, not just the message payload. The `receive` method returns a `Message` object. If the message type is known, then you can cast the return message as we did in the preceding code example. However, a more failsafe technique would be to check the return `Message` type using the `instanceof` keyword as indicated here:

```
Message rmsg = qReceiver.receive(30000);
if (rmsg == null) {
   System.out.println("QLender not responding");
} else {
   if (rmsg instanceof TextMessage) {
   TextMessage tmsg = (TextMessage)rmsg;
      System.out.println("Loan request was " + tmsg.getText());
   } else {
      throw new IllegalStateException("Invalid message type);
   }
}
```

Notice that the message received does not need to be of the same message type as the one sent. In the example just shown, we sent the loan request using a `MapMessage`, yet we received the response from the receiver as a `TextMessage`. While you could potentially increase the level of decoupling between the sender and receiver by including the message type as part of the application properties of the message, you would still need to know how to interpret the payload in the message. For example, with a `StreamMessage` or `BytesMessage` you would still need to know the order of data being sent so that you could in turn read it in the proper order and data type. As you can guess, because of the "contract" of the data between the sender and receiver, there is still a fair amount of coupling in the point-to-point model, at least from the payload perspective.

The QLender Class

The role of the `QLender` class is to listen for loan requests on the loan request queue, determine if the salary meets the necessary business requirements, and finally send the results back to the borrower. Notice that the `QLender` class is structured a bit differently from the `QBorrower` class. In our example, the `QLender` class is referred to as a *message listener* and, as such, implements the `javax.jms.MessageListener` interface and overrides the `onMessage()` method. Here is the complete listing for the `QLender` class:

```
package ch04.p2p;

import java.io.*;
import javax.jms.*;
import javax.naming.*;

public class QLender implements MessageListener {

   private QueueConnection qConnect = null;
   private QueueSession qSession = null;
   private Queue requestQ = null;

   public QLender(String queuecf, String requestQueue) {
      try {
         // Connect to the provider and get the JMS connection
         Context ctx = new InitialContext();
         QueueConnectionFactory qFactory = (QueueConnectionFactory)
            ctx.lookup(queuecf);
         qConnect = qFactory.createQueueConnection();

         // Create the JMS Session
         qSession = qConnect.createQueueSession(
            false, Session.AUTO_ACKNOWLEDGE);

         // Lookup the request queue
         requestQ = (Queue)ctx.lookup(requestQueue);

         // Now that setup is complete, start the Connection
         qConnect.start();

         // Create the message listener
         QueueReceiver qReceiver = qSession.createReceiver(requestQ);
         qReceiver.setMessageListener(this);

         System.out.println("Waiting for loan requests...");

      } catch (JMSException jmse) {
         jmse.printStackTrace();
         System.exit(1);
      } catch (NamingException jne) {
         jne.printStackTrace();
         System.exit(1);
      }
   }

   public void onMessage(Message message) {
      try {
         boolean accepted = false;

         // Get the data from the message
         MapMessage msg = (MapMessage)message;
         double salary = msg.getDouble("Salary");
```

```
            double loanAmt = msg.getDouble("LoanAmount");

            // Determine whether to accept or decline the loan
            if (loanAmt < 200000) {
               accepted = (salary / loanAmt) > .25;
            } else {
               accepted = (salary / loanAmt) > .33;
            }
            System.out.println("" +
               "Percent = " + (salary / loanAmt) + ", loan is "
               + (accepted ? "Accepted!" : "Declined"));

            // Send the results back to the borrower
            TextMessage tmsg = qSession.createTextMessage();
            tmsg.setText(accepted ? "Accepted!" : "Declined");
            tmsg.setJMSCorrelationID(message.getJMSMessageID());

            // Create the sender and send the message
            QueueSender qSender =
               qSession.createSender((Queue)message.getJMSReplyTo());
            qSender.send(tmsg);

            System.out.println("\nWaiting for loan requests...");

         } catch (JMSException jmse) {
            jmse.printStackTrace();
            System.exit(1);
         } catch (Exception jmse) {
            jmse.printStackTrace();
            System.exit(1);
         }
      }

      private void exit() {
         try {
            qConnect.close();
         } catch (JMSException jmse) {
            jmse.printStackTrace();
         }
         System.exit(0);
      }

      public static void main(String argv[]) {
         String queuecf = null;
         String requestq = null;
         if (argv.length == 2) {
            queuecf = argv[0];
            requestq = argv[1];
         } else {
            System.out.println("Invalid arguments. Should be: ");
            System.out.println
               ("java QLender factory request_queue");
            System.exit(0);
         }
```

```
        QLender lender = new QLender(queuecf, requestq);

        try {
           // Run until enter is pressed
           BufferedReader stdin = new BufferedReader
              (new InputStreamReader(System.in));
           System.out.println ("QLender application started");
           System.out.println ("Press enter to quit application");
           stdin.readLine();
           lender.exit();
        } catch (IOException ioe) {
           ioe.printStackTrace();
        }
     }
  }
```

The `QLender` class is what is referred to as an *asynchronous message listener*, meaning that unlike the prior `QBorrower` class it will not block when waiting for messages. This is evident from the fact that the `QLender` class implements the `MessageListener` interface and overrides the `onMessage` method.

The `main` method of the `QLender` class validates the command-line arguments and invokes the constructor by instantiating a new `QLender` class. It then keeps the primary thread alive until the enter key is pressed on the command line.

The constructor in the `QLender` class works much in the same way as the `QBorrower` class. The first part of the constructor establishes a connection to the provider, does a JNDI lookup to get the queue, creates a `QueueSession`, and starts the connection:

```
...
// Connect to the provider and get the JMS connection
Context ctx = new InitialContext();
QueueConnectionFactory qFactory = (QueueConnectionFactory)
   ctx.lookup(queuecf);
qConnect = qFactory.createQueueConnection();

// Create the JMS Session
qSession = qConnect.createQueueSession(
   false, Session.AUTO_ACKNOWLEDGE);

// Lookup the request queue
requestQ = (Queue)ctx.lookup(requestQueue);

// Now that setup is complete, start the Connection
qConnect.start();
...
```

Once the connection is started, the `QLender` class can begin to receive messages. However, before it can receive messages, it must be registered by the `QueueReceiver` as a message listener:

```
QueueReceiver qReceiver = qSession.createReceiver(requestQ);
qReceiver.setMessageListener(this);
```

At this point, a separate listener thread is started. That thread will wait until a message is received, and upon receipt of a message, will invoke the onMessage method of the listener class. In this case, we set the message listener to the QLender class using the this keyword in the setMessageListener method. We could have easily delegated the messaging work to another class that implemented the MessageListener interface:

```
qReceiver.setMessageListener(someOtherClass);
```

When a message is received on the queue specified in the createReceiver method, the listener thread will asynchronously invoke the onMessage method of the listener class (in our case, the QLender class is also the listener class). The onMessage method first casts the message to a MapMessage (the message type we are expecting to receive from the borrower). It then extracts the salary and loan amount requested from the message payload, checks the salary to loan amount ratio, then determines whether to accept or decline the loan request:

```
...
public void onMessage(Message message) {
   try {
      boolean accepted = false;

      // Get the data from the message
      MapMessage msg = (MapMessage)message;
      double salary = msg.getDouble("Salary");
      double loanAmt = msg.getDouble("LoanAmount");

      // Determine whether to accept or decline the loan
      if (loanAmt < 200000) {
         accepted = (salary / loanAmt) > .25;
      } else {
         accepted = (salary / loanAmt) > .33;
      }
      System.out.println("" +
         "Percent = " + (salary / loanAmt) + ", loan is "
         + (accepted ? "Accepted!" : "Declined"));
      ...
```

Again, to make this more failsafe, it would be better to check the JMS message type using the instanceof keyword in the event another message type was being sent to that queue:

```
if (message instanceof MapMessage) {
   //process request
} else {
   throw new IllegalArgumentException("unsupported message type");
}
```

Once the loan request has been analyzed and the results determined, the QLender class needs to send the response back to the borrower. It does this by first creating a JMS message to send. The response message does not need to be the same JMS message type as the loan request message that was received by the QLender. To illustrate this point the QLender returns a TextMessage back to the QBorrower:

```
TextMessage tmsg = qSession.createTextMessage();
tmsg.setText(accepted ? "Accepted!" : "Declined");
```

The next statement sets the `JMSCorrelationID`, which is the JMS header property that is used by the `QBorrower` class to filter incoming response messages:

```
tmsg.setJMSCorrelationID(message.getJMSMessageID());
```

Message correlation is discussed in more detail in the next section of this chapter.

Once the message is created, the `onMessage` method then sends the message to the response queue specified by the `JMSReplyTo` message header property. As you may remember, in the `QBorrower` class we set the `JMSReplyTo` header property when sending the original loan request. The `QLender` class can now use that property as the destination to send the response message to:

```
QueueSender qSender =
   qSession.createSender((Queue)message.getJMSReplyTo());
qSender.send(tmsg);
```

Message Correlation

In the previous code example, the borrower sent a loan request on a request queue and waited for a reply from the lender on a response queue. Many borrowers may be making requests at the same time, meaning that the lender application is sending many messages to the response queue. Since the response queue may contain many messages, how can you be sure that the response you received from the lender was meant for you and not another borrower?

In general, whenever using the request/reply model, you must make sure the response you are receiving is associated with the original message you sent. *Message correlation* is the technique used to ensure that you receive the right message. The most popular method for correlating messages is leveraging the `JMSCorrelationID` message header property in conjunction with the `JMSMessageID` header property. The `JMSCorrelationID` property contains a unique `String` value that is known by both the sender and receiver. The `JMSMessageID` is typically used, since it is unique and is available to the sender and receiver.

When the message consumer (e.g., `QLender`) is ready to send the reply message, it sets the `JMSCorrelationID` message property to the message ID from the original message:

```
public class QLender implements MessageListener {

   ...
   public void onMessage(Message message) {
      try {
         ...
         // Send the results back to the borrower
         TextMessage tmsg = qSession.createTextMessage();
         tmsg.setText(accepted ? "Accepted!" : "Declined");
         tmsg.setJMSCorrelationID(message.getJMSMessageID());
```

```
            // Create the sender and send the message
            QueueSender qSender =
               qSession.createSender((Queue)message.getJMSReplyTo());
            qSender.send(tmsg);

            System.out.println("\nWaiting for loan requests...");
            ...
         }
      }
      ...
```

The original message producer (e.g., `QBorrower`) expecting the response about whether the loan was approved creates a message selector based on the `JMSCorrelationID` message property:

```
   public class QBorrower {

      ...
      private void sendLoanRequest(double salary, double loanAmt) {
         try {
            ...

            // Wait to see if the loan request was accepted or declined
            String filter =
               "JMSCorrelationID = '" + msg.getJMSMessageID() + "'";
            QueueReceiver qReceiver = qSession.createReceiver(responseQ, filter);
            TextMessage tmsg = (TextMessage)qReceiver.receive(30000);
            ...
         }
      }
      ...
```

Although the `JMSMessageID` is typically used to identify the unique message, it certainly is not a requirement. You can use anything that would correlate the request and reply messages. For example, as an alternative you could use the Java `UUID` class to generate a unique ID. In the following code example, the `QBorrower` class generates a unique ID and sets an application message property called "`UUID`" to the generated value:

```
   public class QBorrower {

      ...
      private void sendLoanRequest(double salary, double loanAmt) {
         try {
            // Create JMS message
            MapMessage msg = qSession.createMapMessage();
            msg.setDouble("Salary", salary);
            msg.setDouble("LoanAmount", loanAmt);
            msg.setJMSReplyTo(responseQ);
            UUID uuid = UUID.randomUUID();
            String uniqueId = uuid.toString();
            msg.setStringProperty("UUID", uniqueId);

            // Create the sender and send the message
            QueueSender qSender = qSession.createSender(requestQ);
```

```
            qSender.send(msg);

            // Wait to see if the loan request was accepted or declined
            String filter =
               "JMSCorrelationID = '" + uniqueId + "'";
            QueueReceiver qReceiver = qSession.createReceiver(responseQ, filter);
            TextMessage tmsg = (TextMessage)qReceiver.receive(30000);
            ...
         }
      }
      ...
```

The `QLender` application must now get the `UUID` property from the original message and set the `JMSCorrelationID` message property to this value:

```
   public class QLender implements MessageListener {

      ...
      public void onMessage(Message message) {
         try {
            ...
            // Send the results back to the borrower
            TextMessage tmsg = qSession.createTextMessage();
            tmsg.setText(accepted ? "Accepted!" : "Declined");
            tmsg.setJMSCorrelationID(message.getStringProperty("UUID"));

            // Create the sender and send the message
            QueueSender qSender =
               qSession.createSender((Queue)message.getJMSReplyTo());
            qSender.send(tmsg);

            System.out.println("\nWaiting for loan requests...");
            ...
         }
      }
      ...
```

Although it is commonly used, you are not required to use the `JMSCorrelationID` message header property to correlate messages. As a matter of fact, you could set the correlation property to any application property in the message. While this is certainly possible, you should leverage the header properties if they exist for full compatibility with messaging servers, third-party brokers, and third-party message bridges.

Dynamic Versus Administered Queues

Dynamic queues are queues that are created through the application source code using a vendor-specific API. *Administered queues* are queues that are defined in the JMS provider configuration files or administration tools.

The setup and configuration of dynamic queues tends to be vendor-specific. A queue may be used exclusively by one consumer or shared by multiple consumers. It may have a size limit (limiting the number of unconsumed messages held in the queue) with options for in-memory storage versus overflow to disk. In addition, a queue may be configured with a vendor-specific addressing syntax or special routing capabilities.

JMS does not attempt to define a set of APIs for all the possible options on a queue. It should be possible to set these options administratively, using the vendor-specific administration capabilities. Most vendors supply a command-line administration tool, a graphical administration tool, or an API for administering queues at runtime. Some vendors supply all three. Using vendor-specific administration APIs to create and configure a queue may be convenient at times. However, it is not very portable and may require that the application have administrator privileges.

JMS provides a `QueueSession.createQueue(String queueName)` method, but this is not intended to define a new queue in the messaging system. It is intended to return a `Queue` object that represents an existing queue. There is also a JMS-defined method for creating a temporary queue that can only be consumed by the JMS client that created it using the `QueueSession.createTemporaryQueue()` method.

Creating dynamic queues is useful if you have a large number of queues that may increase over time. For example, consider the scenario where a book publisher has relationships with a large number of bookstores. The book publisher regularly sends new book information and order status to the bookstores. Let's assume that there are 1,000 bookstores related to the book publisher. That equates to 1,000 queues—somewhat excessive to administer. The book publisher can dynamically create the bookstore queues based on a numbering scheme, therefore quickly defining the queues necessary for this scenario (e.g., `BookstoreQ1`, `BookstoreQ2`, etc.).

Load Balancing Using Multiple Receivers

A queue may have multiple receivers attached to it for the purpose of distributing the workload of message processing. The JMS specification states that this capability must be implemented by a JMS provider, although it does not define the rules for how the messages are distributed among consumers. A sender could use this feature to distribute messages to multiple instances of an application, each of which would provide its own receiver.

When multiple receivers are attached to a queue, each message in the queue is delivered to one receiver. The absolute order of messages cannot be guaranteed, since one receiver may process messages faster than another. From the receiver's perspective, the messages it consumes should be in relative order; messages delivered to the queue earlier are consumed first. However, if a message needs to be redelivered due to an acknowledgment failure, it is possible that it could be delivered to another receiver. The other receiver may have already processed more recently delivered messages, which would place the redelivered message out of the original order.

If you would like to see multiple recipients in action, try starting two instances of the QLender class and one instance of the QBorrower class, each in a separate command window:

```
java ch04.p2p.QLender QueueCF LoanRequestQ
java ch04.p2p.QLender QueueCF LoanRequestQ
java ch04.p2p.QBorrower QueueCF LoanRequestQ LoanResponseQ
```

Now, when entering a salary and loan amount in the command window, you will notice that the message is delivered to one or the other QLender application, *but not both*. The exact load balancing scheme will vary between JMS providers. Some may use a round-robin load balancing scheme, whereas others may use a first-available balancing scheme. You will need to consult your JMS provider documentation to determine the specific load balancing algorithm used.

Examining a Queue

A QueueBrowser is a specialized object that allows you to peek ahead at pending messages on a Queue without actually consuming them. This feature is unique to point-to-point messaging. Queue browsing can be useful for monitoring the contents of a queue from an administration tool or for browsing through multiple messages to locate a message that is more important than the one at the head of the queue. It is also useful for other monitoring tasks, such as determining the current queue depth.

Messages obtained from a QueueBrowser are copies of messages contained in the queue and are not considered to be consumed—they are merely for browsing. It is also important to note that the QueueBrowser is not guaranteed to have a definitive list of messages in the queue. The QueueBrowser contains only a snapshot, or a copy of, the queue as it appears at the time the QueueBrowser is created. The contents of the queue may change between the time the browser is created and the time you examine its contents. However, no matter how small that window of time is, new messages may arrive and other messages may be consumed by other JMS clients.

A QueueBrowser is created from the Session object using the createBrowser() method. This method takes as an argument the queue from which you would like to view the messages. It is during the createBrowser() method invocation that the snapshot is taken from the queue. You can then get a list of the messages by using the method getEnumeration() from the QueueBrowser:

```
...
QueueBrowser browser = session.createBrowser(queue);
Enumeration e = browser.getEnumeration();
while (e.hasMoreElements()) {
   //display messages
}
...
```

The full LoanRequestQueueBrowser class is listed here:

```
import java.util.Enumeration;
import javax.jms.Queue;
import javax.jms.QueueBrowser;
import javax.jms.QueueConnection;
import javax.jms.QueueConnectionFactory;
import javax.jms.QueueSession;
import javax.jms.TextMessage;
import javax.naming.Context;
import javax.naming.InitialContext;

public class LoanRequestQueueBrowser {

  public static void main(String[] args) {
    try {
      //establish connection
      Context context = new InitialContext();
      QueueConnectionFactory factory = (QueueConnectionFactory)
         context.lookup("QueueCF");
      QueueConnection connection = factory.createQueueConnection();
      connection.start();

      //establish session
      Queue queue = (Queue) context.lookup("LoanRequestQ");
      QueueSession session = connection.createQueueSession
         (false, Session.AUTO_ACKNOWLEDGE);
      QueueBrowser browser = session.createBrowser(queue);

      Enumeration e = browser.getEnumeration();
      while (e.hasMoreElements()) {
        TextMessage msg = (TextMessage)e.nextElement();
        System.out.println("Browsing: " + msg.getText());
      }

      browser.close();
      connection.close();
      System.exit(0);

    } catch (Exception exception) {
       exception.printStackTrace();
    }
  }
}
```

CHAPTER 5

Publish-and-Subscribe Messaging

This chapter focuses on the publish-and-subscribe (pub/sub) messaging model. This messaging model is used when you need to broadcast an event or message to many message consumers. Unlike the point-to-point messaging model discussed in Chapter 4, all message consumers (called *subscribers*) listening on the topic will receive the message.

In this chapter, we will describe the publish-and-subscribe model through the use of a typical broadcast messaging scenario where a mortgage lender publishes its latest mortgage rates for a 30-year fixed mortgage to various borrowers in the hope of attracting one of the borrowers to apply for a mortgage loan. The `TLender` class will publish a new mortgage rate through a topic, and the `TBorrower` class will subscribe to the topic and make a decision on whether or not it is a good rate.

We have mentioned several new terms already in this chapter: *topic*, *subscriber*, and *publish*. Before moving on to the code example, we will first describe the main characteristics and nomenclature of the publish-and-subscribe model and discuss some of the typical use cases for this model.

Publish-and-Subscribe Overview

The publish-and-subscribe model is more commonly referred to as the pub/sub model. In this model, the message producer is called a *publisher* and the message consumer a *subscriber*. Messages are *published to a topic* as opposed to being sent to a queue, as in the point-to-point model. The most important characteristics of the pub/sub model are as follows:

- Messages are exchanged through a virtual channel called a topic.

- Each message is delivered to multiple message consumers, called subscribers. There are many types of subscribers, including durable, nondurable, and dynamic. These subscriber types are described later in this chapter.
- The publisher generally does not know and is not aware of which subscribers are receiving the topic messages.
- Messages are pushed to consumers, which means that consumers are delivered messages without having to request them. Messages are exchanged through a virtual channel called a *topic*. A topic is a destination where producers can publish, and subscribers can consume, messages. Messages delivered to a topic are automatically pushed to all qualified consumers.
- As in enterprise messaging in general, there is no coupling of the producers to the consumers. Subscribers and publishers can be added dynamically at runtime, which allows the system to grow or shrink in complexity over time.
- Every client that subscribes to a topic receives its own copy of messages published to that topic. A single message produced by one publisher may be copied and distributed to hundreds or even thousands of subscribers.

With the pub/sub model, messages published to a topic are immediately delivered to each subscriber by the JMS provider. Therefore, unlike the point-to-point model, subscribers do not "scan the topic" for messages belonging to them. Rather, the JMS provider delivers a copy of the message to each subscriber.

Another major difference between the pub/sub and p2p models is that, with the pub/sub model, message selectors are applied when the message is copied to each subscriber; whereas with the p2p model, message selectors are applied after the message has been added to the queue.

Subscribers can be either durable or nondurable. Nondurable subscribers receive messages only when that subscriber is currently active and connected to the topic, whereas durable subscribers receive all desired messages sent to that topic, regardless of whether that subscriber is active. Durable and nondurable subscribers are discussed in more detail later in this chapter.

Subscribers may also be dynamic or administered. As you will see later in this chapter, dynamic durable subscribers can be created on the fly, whereas administered subscribers are static and known by the JMS provider. Although the JMS specification allows for the creation of dynamic durable subscribers, there are negative implications associated with this feature which will be discussed later in this chapter. The publish-and-subscribe model is illustrated in Figure 5-1.

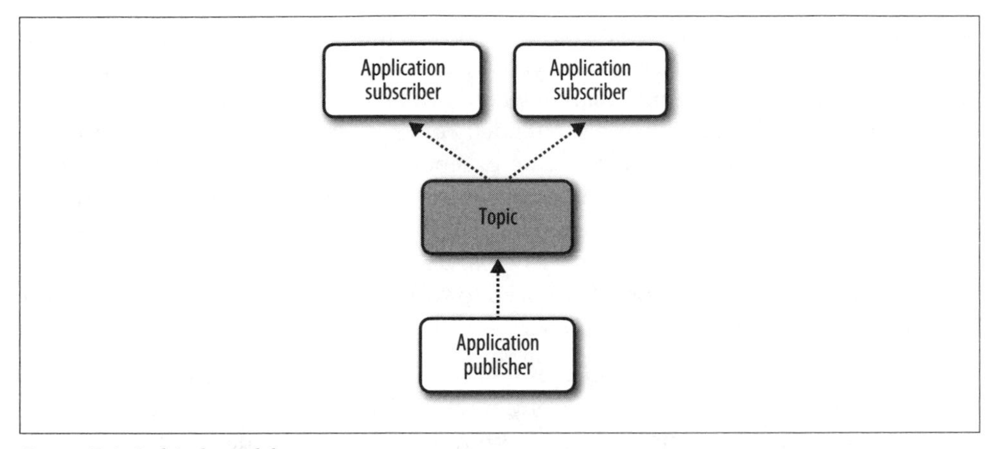

Figure 5-1. Pub/sub model

The specific pub/sub interfaces for connecting, creating, sending, and receiving are shown in Table 5-1.

Table 5-1. Interfaces for topics

General API	Publish-and-subscribe API
`ConnectionFactory`	`TopicConnectionFactory`
`Destination`	`Topic`
`Connection`	`TopicConnection`
`Session`	`TopicSession`
`MessageConsumer`	`TopicPublisher`
`MessageProducer`	`TopicSubscriber`

When to Use Publish-and-Subscribe Messaging

As indicated in Chapter 4, the rationale behind the two models (point-to-point and publish-and-subscribe) lies in the origin of the JMS specification. JMS started out as a way of providing a common API for accessing existing messaging systems. At the time of its conception, some messaging vendors had a p2p model and some had a pub/sub model. Hence, JMS needed to provide an API for both models to gain wide industry support.

The publish-and-subscribe model is used when you want to broadcast a message or event to multiple message consumers. The important point here is that *multiple consumers* may consume the message. By design, the pub/sub model will push copies of the message out to multiple subscribers.

Some of the more common use cases for the pub/sub model are examples such as stock price updates, exception or error notification, and change notification of a particular data item in the database. Any situation where you need to notify multiple consumers of an event is a good use of the pub/sub model. For example, suppose you want to send out a notification to a topic whenever an exception occurs in your application or a system component. You may not know how that information will be used or what type of components will use it. Will the exception be emailed to various parties of interest? Will a notification be sent to a beeper or pager? This is the beauty of the pub/sub model—the publisher does not care or need to worry about how the information will be used; it simply publishes it to a topic.

The same is true of stock price updates—how will that data be used? Is it used for analytics, trend gathering, or to make a buy or sell decision for a particular stock? Again, the publisher of the price quotes does not know or care how the data is used. Its only purpose it to publish the necessary data to the topic, demonstrating the decoupled nature of the pub/sub model.

Conversely, you would not want to use the pub/sub model for such activities as ordering a book or placing a stock trade. In these situations, if the pub/sub model was used, multiple subscribers might pick up the book order or stock trade, resulting in many books delivered to your house or multiple buy or sell orders for the stock. These scenarios would be better suited for the point-to-point model, where the message is guaranteed to be delivered to one and only one consumer.

The TBorrower and TLender Application

To illustrate how pub/sub messaging works, we will use a simple example where a mortgage lender publishes mortgage rates and a borrower interested in the latest rates subscribes to the topic. The lender, implemented through the `TLender` class, will publish a simple `BytesMessage` containing the rate. The borrower, implemented through the `TBorrower` class, will subscribe to the topic through a nondurable subscriber and then decide whether it is a good rate or not.

Since the JMS API for the pub/sub model is similar to the point-to-point model discussed in the previous chapter, we will not be going into as much detail regarding the similar API in this chapter.

Configuring and Running the Application

Before looking at the code, let's look at how the application works. As with the `Chat` application, the `TBorrower` class and `TLender` class both include a `main()` method so they can be run as a standalone Java application. To keep the code vendor-agnostic, both classes need the connection factory name and queue names when starting the application. The `TLender` class is executed from the command line as follows:

```
java ch05.pubsub.TLender ConnectionFactory Topic
```

where ConnectionFactory is the name of the topic connection factory defined in your JMS provider and Topic is the name of the topic where the TLender class publishes the new rate.

The TBorrower class can be executed in the same manner in a separate command window:

```
java ch05.pubsub.TBorrower ConnectionFactory Topic CurrentRate
```

where ConnectionFactory is the name of the topic connection factory defined in your JMS provider, Topic is the name of the topic that the TBorrower class should listen on for updated mortgage rates, and CurrentRate is the current mortgage rate for the borrower.

You will also need to define a *jndi.properties* file in your classpath that contains the JNDI connection information for the JMS provider. The *jndi.properties* file for running the examples in this chapter is similar to the one used in Chapter 4. You will need to consult the documentation of your specific JMS provider or Java EE container to obtain these values. You can find the steps for configuring ActiveMQ to run the examples in this chapter in Appendix D.

The TBorrower and TLender classes both require the topic connection factory name and topic name to run. We have chosen to name the connection factory TopicCF and the loan rate topic RateTopic. These JNDI resources are typically configured in the JMS provider XML configuration files or configuration screens. You will need to consult your JMS provider documentation on how to configure these resources (please refer to Appendix D for the specific configuration settings for ActiveMQ used to run the examples in this chapter).

You can run the TBorrower and TLender classes by entering the following two commands in separate command windows:

```
java ch05.pubsub.TLender TopicCF RateTopic

java ch05.pubsub.TBorrower TopicCF RateTopic 5.6
```

When the TLender class starts, you will be prompted to enter a new mortgage rate. When you press enter, the TLender class will publish the new rate to the topic. The TBorrower class will then receive the new rate and decide whether it is good or not:

```
TLender> TLender Application Started
TLender> Press enter to quit application
TLender> Enter: Rate
TLender> e.g. 6.8

TLender> 6.8
TBorrower> New rate = 6.8 - keep existing loan

TLender> 4.5
TBorrower> New rate = 4.5 - consider refinancing loan
TLender>
```

This interaction is illustrated in Figure 5-2.

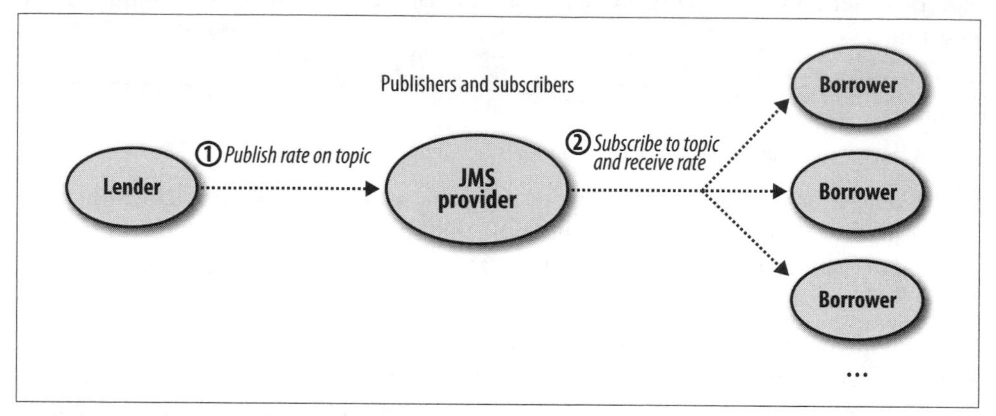

Figure 5-2. Publishers and subscribers

The rest of this chapter examines the source code for the `TBorrower` and `TLender` classes and covers several advanced subjects related to the pub/sub messaging model.

The TLender Class

The `TLender` class is responsible for publishing a new mortgage rate to a topic. The class is fairly straightforward; the constructor establishes a connection to the JMS provider, creates a `TopicSession`, and gets the topic using a JNDI lookup. The `main` method instantiates the `TLender` class and, upon receiving a new rate, invokes the `publishRate` method to publish the message to the topic. Here is the listing for the `TLender` class in its entirety. We will be examining the JMS aspects of this class in detail after the full listing:

```
package ch05.pubsub;

import java.io.*;
import javax.jms.*;
import javax.naming.*;

public class TLender {

    private TopicConnection tConnect = null;
    private TopicSession tSession = null;
    private Topic topic = null;

    public TLender(String topiccf, String topicName) {
       try {
          // Connect to the provider and get the JMS connection
          Context ctx = new InitialContext();
          TopicConnectionFactory qFactory = (TopicConnectionFactory)
             ctx.lookup(topiccf);
          tConnect = qFactory.createTopicConnection();
```

```
            // Create the JMS Session
            tSession = tConnect.createTopicSession(
               false, Session.AUTO_ACKNOWLEDGE);

            // Lookup the request and response queues
            topic = (Topic)ctx.lookup(topicName);

            // Now that setup is complete, start the Connection
            tConnect.start();

         } catch (JMSException jmse) {
            jmse.printStackTrace();
            System.exit(1);
         } catch (NamingException jne) {
            jne.printStackTrace();
            System.exit(1);
         }
      }

      private void publishRate(double newRate) {
         try {
            // Create JMS message
            BytesMessage msg = tSession.createBytesMessage();
            msg.writeDouble(newRate);

            // Create the publisher and publish the message
            TopicPublisher publisher = tSession.createPublisher(topic);
            publisher.publish(msg);

         } catch (JMSException jmse) {
            jmse.printStackTrace();
            System.exit(1);
         }
      }

      private void exit() {
         try {
            tConnect.close();
         } catch (JMSException jmse) {
            jmse.printStackTrace();
         }
         System.exit(0);
      }

      public static void main(String argv[]) {
         String topiccf = null;
         String topicName = null;
         if (argv.length == 2) {
            topiccf = argv[0];
            topicName = argv[1];
         } else {
            System.out.println("Invalid arguments. Should be: ");
            System.out.println("java TLender factory topic");
            System.exit(0);
```

```
         }

         TLender lender = new TLender(topiccf, topicName);

         try {
            // Read all standard input and send it as a message
            BufferedReader stdin = new BufferedReader
               (new InputStreamReader(System.in));
            System.out.println ("TLender Application Started");
            System.out.println ("Press enter to quit application");
            System.out.println ("Enter: Rate");
            System.out.println("\ne.g. 6.8");

            while (true) {
               System.out.print("> ");
               String rate = stdin.readLine();
               if (rate == null || rate.trim().length() <= 0) {
                  lender.exit();
               }

               // Parse the deal description
               double newRate = Double.valueOf(rate);
               lender.publishRate(newRate);
            }
         } catch (IOException ioe) {
            ioe.printStackTrace();
         }
      }
   }
```

The `main` method of the `TLender` class accepts two arguments from the command line: the JNDI name of the topic connection factory and the JNDI name of the topic used to publish the rates. Once the input parameters have been validated, the `TLender` class is instantiated and a loop is started that reads the new mortgage rate from the console:

```
String rate = stdin.readLine();
```

The rate is then parsed and finally the `publishRate` method is invoked. The input loop continues until the user presses enter on the console without entering any data:

```
if (rate == null || rate.trim().length() <= 0) {
   lender.exit();
}
```

Now let's look at the JMS portion of the code in detail, starting with the constructor and ending with the `publishRate` method.

JMS initialization

In the `TLender` class example, all of the JMS initialization logic is handled in the constructor. The code in the `TLender` constructor is almost identical to the `QBorrower` constructor found in Chapter 4 with a couple of important differences. First of all, notice that the connection factory, connection, and session objects are similar to that of the QBorrower class, except that the topic-based interfaces are used instead of the queue-based interfaces:

```
// Connect to the provider and get the JMS connection
Context ctx = new InitialContext();
TopicConnectionFactory qFactory = (TopicConnectionFactory)
   ctx.lookup(topiccf);
tConnect = qFactory.createTopicConnection();

// Create the JMS Session
tSession = tConnect.createTopicSession(
   false, Session.AUTO_ACKNOWLEDGE);

// Lookup the request and response queues
topic = (Topic)ctx.lookup(topicName);

// Now that setup is complete, start the Connection
tConnect.start();
```

The important thing to note here is that although we are using the topic-based API, the flow is the same as that of the queue-based API used with the point-to-point model:

1. Get an initial context to the JMS provider.
2. Look up the connection factory.
3. Create a JMS connection.
4. Create a JMS session.
5. Look up the destination.
6. Start the connection.

Since the details are the same as those for the queue-based API found in Chapter 4, we will not repeat the details of each of these statements.

Publishing the message

Once the TLender class is initialized, the rate is entered through the command line. At this point, the publishRate method is invoked from the main method and the rate published to the topic. Unlike the example in the previous chapter, the TLender class will not wait for a response once the message has been published. This is done intentionally to illustrate the decoupled nature of the pub/sub model; the TLender class does not know or care about who is subscribing to the rates, what they are doing with the data, or how many subscribers are receiving the rate information. There may be subscribers that are receiving the rate data to do some trend analysis on the fluctuation of mortgage rates by this particular lender, whereas other subscribers (like the TBorrower class) are analyzing the rate to determine whether to refinance or not.

At the start of the publishRate method, we create a BytesMessage to hold the rate data. Again, we could have chosen any of the five JMS message types, but we chose the BytesMessage for maximum portability:

```
BytesMessage msg = tSession.createBytesMessage();
msg.writeDouble(newRate);
```

After the message is created, we then create the TopicPublisher object, specifying the topic we wish to publish messages to, and then publish the message using the publish method:

```
// Create the publisher and publish the message
TopicPublisher publisher = tSession.createPublisher(topic);
publisher.publish(msg);
```

Like the send method in the point-to-point model, there are several overridden publish methods available in the TopicSender object. The one we are using here accepts only the JMS message object as the single argument. The other overridden methods allow you to specify the Topic, the delivery mode, the message priority, and finally the message expiry. Since we are not specifying any of the other values in the example just shown, the message priority is set to normal (4), the delivery mode is set to persistent messages (DeliveryMode.PERSISTENT), and the message expiry (time to live) is set to 0, indicating that the message will never expire. All of these parameters can be overridden by using one of the other publish methods.

One note to make at this point: although we are not using the request/reply model in this example, request/reply could certainly apply to the pub/sub model as with the point-to-point messaging model. As a matter of fact, we can publish to a topic and listen for requests on a separate queue. While this is certainly possible, it is not common in today's topic-based messaging models, mostly due to the nature of the pub/sub model. The pub/sub model is generally used to broadcast events or information without expecting a response to that broadcast.

The TBorrower Class

The TBorrower class acts as a subscriber to the rate topic and, as such, is an asynchronous message listener similar to the QLender class found in Chapter 4. Since it is an asynchronous message listener, it implements the javax.jms.MessageListener interface and overrides the onMessage() method. Here is the complete listing for the TBorrower subscriber class:

```
package ch05.pubsub;

import java.io.*;
import javax.jms.*;
import javax.naming.*;

public class TBorrower implements MessageListener {

    private TopicConnection tConnect = null;
    private TopicSession tSession = null;
    private Topic topic = null;
    private double currentRate;
```

```
public TBorrower(String topiccf, String topicName, String rate) {
   try {
      currentRate = Double.valueOf(rate);

      // Connect to the provider and get the JMS connection
      Context ctx = new InitialContext();
      TopicConnectionFactory qFactory = (TopicConnectionFactory)
         ctx.lookup(topiccf);
      tConnect = qFactory.createTopicConnection();

      // Create the JMS Session
      tSession = tConnect.createTopicSession(
         false, Session.AUTO_ACKNOWLEDGE);

      // Lookup the request and response queues
      topic = (Topic)ctx.lookup(topicName);

      // Create the message listener
      TopicSubscriber subscriber = tSession.createSubscriber(topic);
      subscriber.setMessageListener(this);

      // Now that setup is complete, start the Connection
      tConnect.start();

      System.out.println("Waiting for loan requests...");

   } catch (JMSException jmse) {
      jmse.printStackTrace();
      System.exit(1);
   } catch (NamingException jne) {
      jne.printStackTrace();
      System.exit(1);
   }
}

public void onMessage(Message message) {
   try {
      // Get the data from the message
      BytesMessage msg = (BytesMessage)message;
      double newRate = msg.readDouble();

      // If the rate is at least 1 point lower than the current rate, then
      //recommend refinancing
      if ((currentRate - newRate) >= 1.0) {
         System.out.println(
            "New rate = " + newRate + " - Consider refinancing loan");
      } else {
         System.out.println("New rate = " + newRate + " - Keep existing loan");
      }

      System.out.println("\nWaiting for rate updates...");

   } catch (JMSException jmse) {
      jmse.printStackTrace();
      System.exit(1);
```

```
            } catch (Exception jmse) {
                jmse.printStackTrace();
                System.exit(1);
            }
        }

        private void exit() {
            try {
                tConnect.close();
            } catch (JMSException jmse) {
                jmse.printStackTrace();
            }
            System.exit(0);
        }

        public static void main(String argv[]) {
            String topiccf = null;
            String topicName = null;
            String rate = null;
            if (argv.length == 3) {
                topiccf = argv[0];
                topicName = argv[1];
                rate = argv[2];
            } else {
                System.out.println("Invalid arguments. Should be: ");
                System.out.println("java TBorrower factory topic rate");
                System.exit(0);
            }

            TBorrower borrower = new TBorrower(topiccf, topicName, rate);

            try {
                // Run until enter is pressed
                BufferedReader stdin = new BufferedReader
                    (new InputStreamReader(System.in));
                System.out.println ("TBorrower application started");
                System.out.println ("Press enter to quit application");
                stdin.readLine();
                borrower.exit();
            } catch (IOException ioe) {
                ioe.printStackTrace();
            }
        }
    }
```

The `main` method of the `TBorrower` class validates the command-line arguments and invokes the constructor by instantiating a new `TBorrower` class. It then keeps the primary thread alive until the Enter key is pressed on the command line.

The constructor in the `TBorrower` class works much in the same way as the `TLender` class. The first part of the constructor establishes a connection to the provider, does a JNDI lookup to get the topic, creates a `TopicSession`, and starts the connection:

```
...
// Connect to the provider and get the JMS connection
Context ctx = new InitialContext();
TopicConnectionFactory qFactory = (TopicConnectionFactory)
   ctx.lookup(topiccf);
tConnect = qFactory.createTopicConnection();

// Create the JMS Session
tSession = tConnect.createTopicSession(
   false, Session.AUTO_ACKNOWLEDGE);

// Lookup the request and response queues
topic = (Topic)ctx.lookup(topicName);
...
```

Once the connection is started, the `TBorrower` class can begin to receive messages. However, before it can receive messages, it must be registered by the `TopicSubscriber` as a message listener (in this case, a subscriber):

```
TopicSubscriber subscriber = tSession.createSubscriber(topic);
subscriber.setMessageListener(this);
```

At this point, a separate listener thread is started. That thread will wait until a message is received, and upon receipt of a message will invoke the `onMessage` method of the listener class. In this case, we set the message listener to the `TBorrower` object using the `this` keyword in the `setMessageListener` method.

When a message is received on the topic specified in the `createSubscriber` method, the listener thread will asynchronously invoke the `onMessage` method of the listener class (in our case the `TBorrower` class is also the listener class). The `onMessage` method first casts the message to a `BytesMessage` (the message type we are expecting to receive from the lender). It then extracts the new rate and determines whether to refinance or not:

```
...
public void onMessage(Message message) {
   try {
      // Get the data from the message
      BytesMessage msg = (BytesMessage)message;
      double newRate = msg.readDouble();

      // If the rate is at least 1 point lower than the current rate, then
      //recommend refinancing
      if ((currentRate - newRate) >= 1.0) {
         System.out.println("New rate = " + newRate + " - Consider refinancing loan");
      } else {
         System.out.println("New rate = " + newRate + " - Keep existing loan");
      }
      ...
```

In practice, it would be better to make this method more failsafe by checking the JMS message type using the `instanceof` keyword in the event another message type was being sent to that queue:

```
if (message instanceof BytesMessage) {
   //process request
} else {
   throw new IllegalArgumentException("unsupported message type");
}
```

Durable Versus Nondurable Subscribers

If you were to run the TBorrower class and then publish several rates, the TBorrower class would pick up the new rate and make a determination as to whether it was a good rate or not. However, if you were to terminate the TBorrower class, publish some new rates, then restart the TBorrower class, you would not have picked up the rates that were published to the topic while the TBorrower class was not running. Why? Because the TBorrower class was created as a nondurable subscriber:

```
TopicSubscriber subscriber = tSession.createSubscriber(topic);
```

Nondurable subscribers receive messages only when they are actively listening on that topic. Otherwise, the message is gone. In the pub/sub model, there is no real concept of a "topic" holding all of the messages; rather, when a message is received by the JMS provider, the provider makes a copy of that message for each subscriber. If the subscriber is not active, it does not receive a copy of that message. This concept is illustrated in Figure 5-3.

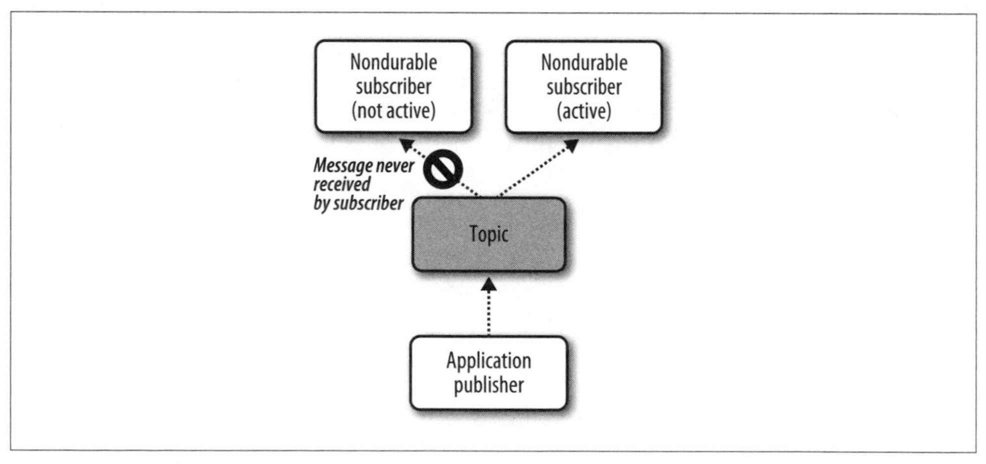

Figure 5-3. Nondurable subscribers

Durable subscribers, on the other hand, will receive all messages sent to that topic (depending on the message selectors applied to that subscriber), regardless of whether that subscriber is active or not. This is commonly referred to as "store-and-forward" messaging. Email is a good example of this concept, even though email is usually not implemented using messaging. You may receive an email in the middle of the night when your computer is off and you are asleep. When you turn your computer on in the morning or when you get to work, sure enough, you receive the email, even through you were not actively connected to the email provider. Figure 5-4 illustrates the store-and-forward concept of durable subscribers.

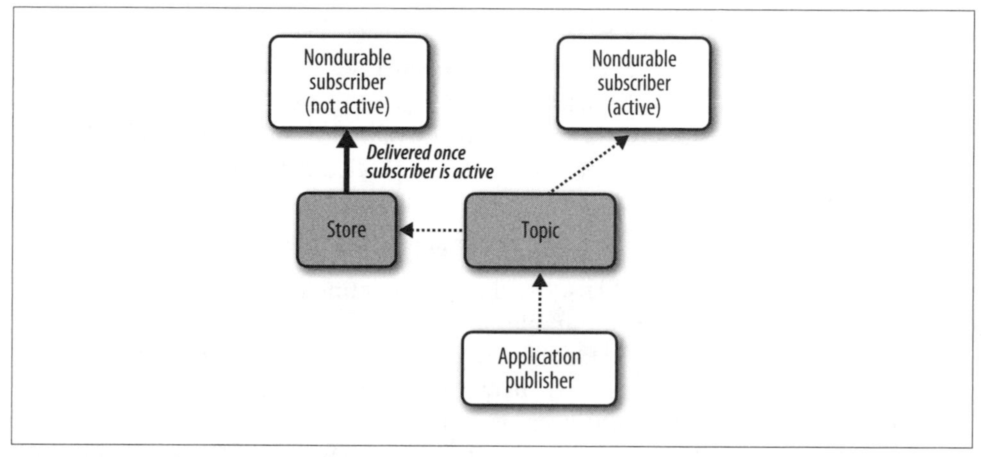

Figure 5-4. Durable subscribers

Durable subscribers are created by specifying the subscriber name in the JMS provider (either through configuration or through an admin interface) and using the method `createDurableSubscriber`, which accepts a subscription name as one of the parameters:

```
TopicSubscriber subscriber = tSession.createDurableSubscriber(topic, "Borrower1");
```

There are many cases where a durable subscriber makes sense and other times when it does not. Although business need generally dictates the choice between a durable and nondurable subscriber, there are several other considerations to take into account, including the volatility of the data and the amount of storage consumed by the messages. For example, stock price updates published every 30 seconds is probably not a good use case for a durable subscriber. First, unless you are doing some sort of trend analysis, generally you would be more concerned about the current price of a stock, not the price 20 or 30 minutes ago. Second, if the durable subscriber is inactive for a long period of time, thousands of useless messages will be stored for that subscriber, wasting valuable space in the JMS datastore. These considerations should be taken into account when deciding between durable and nondurable subscribers.

Dynamic Versus Administered Subscribers

In the previous section, we created a durable subscriber named `Borrower1`:

```
TopicSubscriber subscriber = tSession.createDurableSubscriber(topic, "Borrower1");
```

Some JMS providers allow you to statically define the durable subscriber in the configuration file or admin interface. In this case, the durable subscriber is said to be an *administered durable subscriber*, meaning that the durable subscriber is statically defined and known by the JMS provider. However, suppose you needed to produce a temporary durable subscriber, say to gather mortgage rates for the next one or two days to do some trend analysis. It would be silly to have to modify the JMS provider configuration files for this simple request.

The JMS specification allows for durable subscribers to be defined dynamically at runtime, without having to statically define them in your JMS provider configuration files. These types of durable subscribers are known as *dynamic durable subscribers*. For example, if we were to define a new durable subscriber called `BorrowerA`, we could simply do the following:

```
TopicSubscriber subscriber = tSession.createDurableSubscriber(topic, "BorrowerA");
```

In this case, the `BorrowerA` durable subscriber is not defined in the JMS provider and, therefore, is not an administered durable subscriber. However, once the line of code listed above executes, a new durable subscriber called `BorrowerA` is created in the JMS provider and, therefore, will receive all rates published to the topic, whether the subscriber is active or not. The subscriber will remain a durable subscriber until it is unsubscribed (see the section "Unsubscribing Dynamic Durable Subscribers" on page 104).

Although this feature provides a great deal of flexibility, it also comes with a price. Each durable subscriber, whether it is administered or dynamic, will receive a copy of the message published to the topic.*

This means that when the subscriber is not active, those messages are being stored for each durable subscriber. From a capacity planning standpoint, dynamic durable subscribers are somewhat dangerous in that it is difficult to control the number of durable subscribers using the system (although this can be monitored through an admin console, depending on the JMS provider and monitoring software you are using). Imagine for a moment that 100 new dynamic durable subscribers were suddenly created to start receiving every mortgage rate or stock price to perform trend analysis. Then, once that analysis was complete, those 100 subscribers were retired, but not unsubscribed. This means that every mortgage rate and every stock price update would be stored for those retired dynamic durable subscribers forever or until the machine hosting the JMS datastore ran out of storage or memory.

There are a few methods a middleware administrator can use for addressing this issue in production environments to help control machine resources and capacity. You can prohibit dynamic durable subscribers in your messaging system by frequently (once a minute, once an hour, etc.) running a control program or database script that compares the known durable subscribers with those registered with the JMS provider. Each JMS provider will store the messages in either a database or filesystem. For example,

* At least theoretically. Each JMS provider may have a different implementation of how messages are associated with durable subscribers.

OpenJMS—an open source JMS provider useful for testing and training purposes—uses a JDBC 2.0-compliant database to store messages. If you look at the MySQL database schema in OpenJMS,[†] you will observe two tables of interest—the `consumers` table and the `message_handles` table:

```
create table consumers (
   name varchar(255) not null,
   destinationId bigint not null,
   consumerId bigint not null,
   created bigint not null );

create unique index consumers_pk on consumers(
   name, destinationId);

create table message_handles (
   messageId varchar(64) not null,
   destinationId bigint not null,
   consumerId bigint not null,
   priority int,
   acceptedTime bigint,
   sequenceNumber bigInt,
   expiryTime bigint,
   delivered int );

create index message_handles_pk ON message_handles(
   destinationId, consumerId, messageId);
```

For pub/sub messaging, the `consumers` table is used to hold durable subscribers and the `message_handles` table is used to link the messages to the consumers. Given this schema, a middleware administrator can write a simple database script or program to query for any durable subscribers in the `consumer` table that are not in the administered list of subscribers, and simply delete them from the JMS provider database (along with the corresponding messages in the `message_handles` table).

Another solution is to allow for the creation of dynamic durable subscribers, but only have them active for a limited time period (e.g., two days, one week, etc.). If you notice in the previous MySQL database schema definitions for OpenJMS, the `consumers` table has a `created` column containing the timestamp (represented as a `long` in milliseconds) of when that durable subscriber was first created. You can easily create a database script or control program that executes each evening, removing any dynamic durable subscribers that were created a specified number of days ago. With this method, you can allow for the flexibility of dynamic durable subscribers if the business rules or use cases call for them, but limit the lifespan of those dynamic durable subscribers to avoid filling up the storage capacity of the database.

A less aggressive approach would be to leverage the database schema of the JMS provider to create a report of the number of dynamic durable subscribers and current

[†] OpenJMS is a simple open source JMS provider useful for testing and training purposes. It can be found at *http://openjms.sourceforge.net*.

message count using the tables described earlier. This report would show any significantly large message count for a particular subscriber, indicating that the durable subscriber is possibly retired or no longer interested in the data. The dynamic subscriber would then be flagged as a possible candidate for removal or message cleanup.

Unsubscribing Dynamic Durable Subscribers

There may be cases where you want to explicitly unsubscribe a durable subscriber in a client application. To remove a dynamic durable subscription, you can invoke the `Session.unsubscribe` method:

```
...
private void exit() {
   try {
      subscriber.close();
      tSession.unsubscribe("BorrowerA");
      tConnect.close();
   } catch (javax.jms.JMSException jmse){
      jmse.printStackTrace();
   }
   System.exit(0);
}
...
```

For nondurable subscriptions, calling the `close()` method on the `TopicSubscriber` class is sufficient. For durable subscriptions, there is an `unsubscribe(String name)` method on the `TopicSession` object, which takes the subscription name as its parameter. This informs the JMS provider that it should no longer store messages on behalf of this client. You cannot call the `unsubscribe()` method without first closing the subscription (you will get an exception if you do this). Hence, both methods need to be called for durable subscriptions.

Temporary Topics

In the chat example we explored in Chapter 2, we assumed that JMS clients would communicate with each other using established topics on which messages are asynchronously produced and consumed. In this section, we will explore ways to augment this basic mechanism by looking at temporary topics, which is a mechanism for JMS clients to create topics dynamically.

A temporary topic is a topic that is dynamically created by the JMS provider, using the `createTemporaryTopic` method of the `TopicSession` object. A temporary topic is associated with the connection that belongs to the `TopicSession` that created it. It is only active for the duration of the connection and it is guaranteed to be unique across all connections. Since it is temporary, it can't be durable—it lasts only as long as its associated client connection is active. In all other respects, it is just like a "regular" topic.

Since a temporary topic is unique across all client connections (it is obtained dynamically through a method call on a client's session object), it is unavailable to other JMS clients unless the topic identity is transferred using the `JMSReplyTo` header. While any client may publish messages on another client's temporary topic, only the sessions that are associated with the JMS client connection that created the temporary topic may subscribe to it. JMS clients can also, of course, publish messages to their own temporary topics.

A temporary topic allows a consumer to respond directly to a producer. In larger real-world applications, however, there may be many publishers and subscribers exchanging messages across many topics. A message may represent a workflow, which may take multiple hops through various stages of a business process. In that type of scenario, the consumer of a message may never respond directly to the producer that originated the message. It is more likely that the response to the message will be forwarded to some other process. Thus, the `JMSReplyTo` header can be used as a place to specify a forwarding address, rather than the destination address of the original sender.

CHAPTER 6
Message Filtering

There may be times when you want to be more selective about the messages received from a particular queue or topic. Without message filtering, topic subscribers receive *every* message published to the topic and queue receivers receive the next available message, regardless of the message content or type. In the case of topic subscribers, the subscriber may be forced to process a large number of unnecessary and unwanted messages, usually leading to custom-written Java code to manually filter unwanted messages. A good example of this is with the `TBorrower` class from the prior chapter. In this case, the `TBorrower` class is receiving *every* loan rate published from the `TLender` on the `RateTopic` topic, and then using conditional logic to determine whether to refinance the existing mortgage loan:

```
public class TBorrower implements javax.jms.MessageListener {

    ...
    public TBorrower(String topiccf, String topicName, String rate) {
        try {
            ...
            TopicSubscriber subscriber =
                session.createSubscriber();
            ...
    }

    public void onMessage(Message message) {
       try {
          // Get the data from the message
          BytesMessage msg = (BytesMessage)message;
          double newRate = msg.readDouble();

          // If the rate is at least 1 point lower than the current rate, then
          //recommend refinancing
          if ((currentRate - newRate) >= 1.0) {
             System.out.println(
                 "New rate = " + newRate + " - Consider refinancing loan");
          } else {
             System.out.println("New rate = " + newRate + " - Keep existing loan");
          }
```

```
            System.out.println("\nWaiting for rate updates...");
        ...
    }

    ...
}
```

In this example, the TBorrower subscriber may process thousands of messages before finding one that is a good deal, unnecessarily consuming precious machine resources (e.g., memory and CPU) in the process. A better approach in this case would be to use message filtering so that the subscriber only receives messages that it deems good deals, thereby making the TBorrower subscriber processing more efficient:

```
public class TBorrower implements javax.jms.MessageListener {

    ...
    public TBorrower(String topiccf, String topicName, String rate) {
        try {
            ...
            topic = (Topic)ctx.lookup(topicName);
            ...
            String filter = "(currentRate - newRate) >= 1.0";
            TopicSubscriber subscriber =
                session.createSubscriber(topic, filter, true);
            ...
    }

    public void onMessage(Message message) {
       try {
          // Get the data from the message
          BytesMessage msg = (BytesMessage)message;
          double newRate = msg.readDouble();

          //we received a good deal, start the refinancing process
          ...

          System.out.println("\nWaiting for rate updates...");
       ...
    }
    ...
}
```

This example is only meant to illustrate the possibilities of message filtering. For the code to work, the currentRate and newRate would both have to be exposed as message properties by the message sender (you cannot use local variables or data in the message body as part of a message selector). Notice that the TBorrower subscriber still has control over the rules for what it considers a good deal (specified when creating the subscriber). Also notice that since every message received from the TBorrower subscriber is now a good deal, the onMessage method is no longer required to have conditional logic applied to every message to determine whether it should automatically refinance an existing loan. By using message selectors, the number of messages it receives and processes is significantly reduced, making the overall system processing much more efficient.

Message filtering with queues is a much more interesting case because, unlike topics, once a message is consumed by one queue receiver, it is no longer available to any other queue receiver. This means that if one queue receiver consumes a message and decides it shouldn't process it, it is too late; the message has already been received and will be removed from the queue. For example, assume we have a single queue that holds orders from a retailer. Depending on the status of the customer, orders are either processed with a high priority or normal priority. In this case, two different classes handle the processing of high priority and normal priority orders (`PriorityHandling` class and `NormalHandling` class, respectively). Message filtering based on the customer status would be required since both the `PriorityHandling` and `NormalHandling` classes are both receiving messages on the same queue. If message filtering was not used, the `NormalHandling` class might consume a message for a customer with a status eligible for high-priority order handling, which would in turn require the `NormalHandling` class to somehow get the message to the `PriorityHandling` class (e.g., through inter-service communication or resending the message).

When a JMS consumer declares a message selector for a particular destination, the selector is applied only to messages delivered to that consumer. Every JMS client can have a different selector specified for each of its consumers.

Filtering out certain messages from a queue and/or topic is done through *message selectors*. This chapter describes the specification, use cases, and design considerations for using message selectors within JMS.

Message Selectors

Message selectors are applied to message consumers when creating a `QueueReceiver`, `QueueBrowser`, or `TopicSubscriber`. When message selectors are used, the consumer will receive only messages that apply to the specified filter. Message selectors use message properties and headers as criteria in conditional expressions. These conditional expressions use boolean logic to declare which messages should be delivered to a JMS consumer. Note that a message selector cannot refer to data *within* the message body; only message header and message properties can be used. This means that the message producer must add the appropriate data to the message properties area of the message so that messages can be logically filtered.

Message selectors are based on a subset of the SQL-92 conditional expression syntax. If you are familiar with SQL-92, the conditional expressions used in message selectors will be familiar to you. Message selectors are made up of three elements: identifiers, literals, and comparison operators. The details about each of these elements are described in the following sections.

To illustrate how message selectors are applied, in the following sections we will consider a hypothetical stock trading message that contains three application properties: `Symbol`, `Side`, and `Shares`. `Symbol` is a `String` property containing the stock symbol; `Side` is a `String` property indicating whether this is a buy or sell order; and `Shares` is a `double` property indicating the number of shares to be purchased or sold. The values of these properties depend on the message. The message selector is used to obtain only those messages with property values of interest to the consumer.

Identifiers

An identifier is the part of the expression that is being compared. Identifiers must come from either the message header or message properties. For example, the identifiers in the following expression are `Symbol`, `Side`, `Shares`, and `JMSPriority`:

```
Symbol = 'ABC' AND Side = 'BUY' AND Shares <= 1000.0 AND JMSPriority > 4
```

Identifiers can be any application-defined, JMS-defined, or provider-specific property, or one of several JMS headers. In the example just shown `Symbol`, `Side`, and `Shares` come from application properties in the message, whereas the `JMSPriority` comes from the message header. Identifiers are case-sensitive and must match the property or JMS header name exactly. Identifiers have the same naming restrictions as property names (see Appendix C). Thus, to use the identifiers above, the message producer must set the `Symbol`, `Side`, and `Shares` properties prior to sending the message:

```
...
ObjectMessage msg = session.createObjectMessage(tradeOrder);
msg.setStringProperty("Symbol", tradeOrder.getSymbol());
msg.setStringProperty("Side", tradeOrder.getSide());
msg.setDoubleProperty("Shares", tradeOrder.getShares());
...
```

The JMS headers that can be used as identifiers include the following:

- `JMSDeliveryMode`
- `JMSPriority`
- `JMSMessageID`
- `JMSTimestamp`
- `JMSCorrelationID`
- `JMSType`

The `JMSDestination` and `JMSReplyTo` headers cannot be used as identifiers because their corresponding values are `Destination` objects whose underlying value is proprietary and therefore undefined. The `JMSRedelivered` value may be changed during delivery and is therefore not allowed in a selector. If a consumer uses a message selector where `"JMSRedelivered = false"` and there was a failure delivering a message, then the `JMSRedelivered` flag might be set to `true`. `JMSExpiration` is not supported as an identifier because JMS providers may choose to implement this value differently. Some may store it with the message, while others calculate it as needed.

Literals

Literals are expression values that are hardcoded into the message selector. In the message selector shown here, 'ABC', 'SELL', and 1000 are all literals:

```
Symbol = 'ABC' AND Side = 'BUY' AND Shares <= 1000.0
```

String literals are enclosed in single quotes. An apostrophe or single quote can be included in a String literal by using two single quotes (e.g., 'Smith''s').

Numeric literals are expressed using exact numerical (+22, 30, -52134), approximate numerical with decimal (-33.22, 100.00, +7.0) or scientific (-9E4, 3.5E6) notation.

Boolean literals are expressed as true or false.

Comparison Operators

Comparison operators compare identifiers to literals in a boolean expression that evaluates to either true or false. Comparison operations can be combined into more complex expressions using the logical operators AND and OR. The comparison operators that are used with message selectors include:

- Algebraic comparison operators
- LIKE operator
- BETWEEN operator
- IN operator
- NOT operator
- IS NULL operator

Message selector expressions are evaluated from left to right:

```
Symbol = 'ABC' AND Side = 'BUY' OR Shares <= 1000.0
```

In this example, the expression would be evaluated as if it had parentheses placed as follows (parentheses can be used to group expressions and can change the precedence of evaluation):

```
(Symbol = 'ABC' AND Side = 'BUY') OR Shares <= 1000.0
```

Either the Shares must be less than or equal to 1000.0 or the Shares can be any value as long as the Symbol equals 'ABC' and the Side equals 'BUY'. Evaluating these kinds of expressions should be second nature for most programmers.

Message selectors support six algebraic comparison operators, which are = , > , >= , < , <= , and <> (not equal). These algebraic comparison operators can be used on any of the primitive property types except for **boolean**. The **boolean** and **String** property types are restricted to the = or the <> algebraic operators. A mismatch between the identifier type and the operations allowed on that type will result in an **InvalidSelectorException**.

String types can be compared using the `LIKE` comparison operator. For example:

```
Shares > 1000.0 AND Symbol LIKE 'A%C'
```

The `LIKE` comparison operator attempts to match each character in the literal with characters of the property value. Two special wildcard characters, underscore (_) and percent (%), can be used with the `LIKE` comparison. The underscore stands for any single character. The percent symbol stands for any sequence of characters. All other characters stand for themselves and are case sensitive. Table 6-1 provides some examples of successful and unsuccessful comparisons using the `LIKE` operator.

Table 6-1. Comparisons using the LIKE operator

Expression	True for values	False for values
`LName LIKE 'A_C'`	A**B**C, A**E**C, A**Z**C	A**BQ**C, ABD, AB
`LName LIKE 'AB_'`	AB**C**, AB**Q**, AB**Z**	A**Q**C, ABC**D**
`LName LIKE 'A%C'`	A**B**C, A**SF**C, AC	A**Q**, **B**FTC, AC**D**
`LName LIKE '%CD'`	**AB**CD, **Q**CD, CD	AB**Q**D, ACD**X**

The `BETWEEN` operator can be used to specify a range (inclusive). For example:

```
Shares BETWEEN 1000 and 2000
```

This expression is the same as:

```
(Shares >= 1000) AND (Shares <= 2000)
```

The `IN` operator can be used to specify membership in a set:

```
Symbol IN ('ABC', 'AQC', 'BCD')
```

This expression is the same as:

```
(Symbol = 'ABC') OR (Symbol = 'AQC') OR (Symbol = 'BCD')
```

The `NOT` logical operator can be used in combination with the `LIKE`, `BETWEEN`, `IN`, and `IS NULL` (discussed later) operators to reverse their evaluation. If the expression would have evaluated to **true**, it becomes **false**, and vice versa.

When no property or header exists to match an identifier in a message selector, the value of the identifier is assigned a `null` value. Nonexistent properties evaluating to `null` present some problems with message selectors. In some cases, the `null` value of the property cannot be evaluated in a conditional expression. The result is an *unknown* evaluation—a nice way of saying the result is not predictable across JMS providers. If, for example, a particular message contains the `Symbol` and `Side` properties but does not have a `Shares` property, then the message selector following would evaluate to *unknown* as shown:

```
Symbol = 'ABC' AND Side = 'BUY' OR Shares >= 1000.0

  ____         _____      _______
  TRUE   AND   FALSE  OR  UNKNOWN
```

The results of evaluating unknown expressions with logical operators (AND, OR, NOT) are shown in Tables 6-2 through 6-4.

Table 6-2. Definition of the AND operator

Expression	Result
TRUE AND TRUE	TRUE
TRUE AND FALSE	FALSE
TRUE AND Unknown	Unknown
FALSE AND Unknown	FALSE
Unknown AND Unknown	Unknown

Table 6-3. Definition of the OR operator

Expression	Result
TRUE OR TRUE	TRUE
TRUE OR FALSE	TRUE
TRUE OR Unknown	TRUE
FALSE OR Unknown	Unknown
Unknown OR Unknown	Unknown

Table 6-4. Definition of the NOT operator

Expression	Result
NOT TRUE	FALSE
NOT FALSE	TRUE
NOT Unknown	Unknown

To avoid problems, the `IS NULL` or `IS NOT NULL` comparison can be used to check for the existence of a property:

```
Shares IS NULL AND Symbol IS NOT NULL
```

This expression selects messages that do not have a `Shares` property but do have a `Symbol` property.

Arithmetic Operators

In addition to normal comparison operators, message selectors can use arithmetic operators to calculate values for evaluation dynamically at runtime. Table 6-5 shows the arithmetic operators in their order of precedence.

Table 6-5. Arithmetic operators

Type	Symbol
Unary	+, -
Multiplication and division	*, /
Addition and subtraction	+, -

For example, the following expression applies arithmetic operations to the `Price` and `Shares` properties of a message to only select trade messages in excess of one million dollars:

```
(Price * Shares) > 1000000.00
```

Declaring a Message Selector

When a consumer is created with a message selector, the JMS provider must validate that the selector statement is syntactically correct. If the selector is not correct, the operation throws a `javax.jms.InvalidSelectorException`. For the point-to-point model, message selectors can be applied to a `QueueBrowser` and the `QueueReceiver`, specifically within the `createBrowser()` and `createReceiver()` methods of the `QueueSession`:

```
public interface QueueSession extends Session {

  public QueueBrowser createBrowser(Queue queue,
                                    String messageSelector)
     throws JMSException,
            InvalidSelectorException,
            InvalidDestinationException;

  public QueueReceiver createReceiver(Queue queue,
                                      String messageSelector)
     throws JMSException,
            InvalidSelectorException,
            InvalidDestinationException;
  ...
}
```

For the publish-and-subscribe model, message selectors can be applied to a durable or nondurable `TopicSubscriber` (topic browsing is not supported for the publish-and-subscribe model). When creating a topic subscriber, you can specify the message selector in the `createSubscriber()` or `createDurableSubscriber()` methods of the `TopicSession`:

```
public interface TopicSession extends Session {

   public TopicSubscriber createSubscriber(Topic topic,
                                           String messageSelector,
                                           boolean noLocal)
      throws JMSException,
```

```
                InvalidSelectorException,
                InvalidDestinationException;

    public TopicSubscriber createDurableSubscriber(Queue queue,
                                                   String name,
                                                   String messageSelector,
                                                   boolean noLocal)
        throws JMSException,
               InvalidSelectorException,
               InvalidDestinationException;
    ...
}
```

Notice in the publish-and-subscribe model, when specifying the message selector when creating a subscriber, you must also specify a boolean value for the `noLocal` argument. The `noLocal` argument is only applicable for topics and specifies whether messages published from this message producer should be delivered to this message producer. A value of `true` inhibits messages from being delivered to the same connection that published those messages.

Because the general `Session` interface applies to both the point-to-point and publish-and-subscribe models (and the corresponding `QueueSession` and `TopicSession` interfaces), you can also use the `Session` interface to create a generic `MessageConsumer` using the `createConsumer()` method, specifying the message selector as follows:

```
public interface Session extends Runnable {

  public MessageConsumer createConsumer(Destination dest,
                                        String messageSelector)
     throws JMSException,
            InvalidSelectorException,
            InvalidDestinationException;

  public MessageConsumer createConsumer(Destination dest,
                                        String messageSelector,
                                        boolean noLocal)
     throws JMSException,
            InvalidSelectorException,
            InvalidDestinationException;
  ...
}
```

You can specify the message selector as a string value directly in the method to create the message consumer or you can use a separate `String` variable defined outside the scope of the method call. Specifying a null or empty string value in the message selector indicates that no message selector is to be used for this message consumer.

The message selector used for a consumer can always be obtained by calling the `getMessageSelector()` method on a `QueueReceiver`, `QueueBrowser`, or `TopicSubscriber`. The `getMessageSelector()` method returns the message selector for that consumer as a `String`.

Once a consumer's message selector has been established, it cannot be changed while that message consumer is active. To change a message selector, you must first close the active message consumer and recreate it using the new message selector.

Message Selector Examples

The following are four selectors used in hypothetical environments. Although you will have to use a little imagination, the purpose of these examples is to convey the power of the message selectors. Notice in these examples that a message selector can be applied to both the publish-and-subscribe model and the point-to-point model.

Managing Claims in an HMO

Due to some fraudulent claims, an automatic process is implemented that will audit all claims submitted by patients who are employees of the ACME manufacturing company with visits to chiropractors, psychologists, and dermatologists:

```
String selector =
  "PhysicianType IN ('Chiropractor', 'Psychologist', 'Dermatologist') "
            + "AND PatientGroupID LIKE 'ACME%'";

TopicSubscriber subscriber =
   session.createSubscriber(topic, selector, false);
```

Notification of Certain Bids on Inventory

A supplier wants notification of requests for bids on specific inventory items at specific quantities:

```
String selector =
  "InventoryID = 'S93740283-02' AND Quantity BETWEEN 1000 AND 13000";

TopicSubscriber subscriber =
   session.createSubscriber(topic, selector, false);
```

Priority Handling

A supplier receiving orders handles two types of customers: gold and silver. Since the priority and handing of orders differs greatly between these customer types, the online supplier has different processes to handle gold customers and silver customers, even though all orders are sent to the same queue. Priority handling is offered to gold customers or those messages with a high priority (notice here the use of both a header property and an application-specified property in the message selector):

```
String selector = "CustomerType = 'GOLD' OR JMSPriority BETWEEN 5 AND 9";

QueueReceiver receiver = session.createReceiver(queue, selector);
```

Stock Trade Order Auditing

As part of a standard stock trade order process, whenever a stock trade order is placed, a corresponding message is published on a topic containing the details of that trade. An audit process subscribes to the trade order topic to ensure compliance with the allowable commission rates. For trade orders less than one million dollars, the commission rate is 2.3%. For trade orders greater than or equal to one million dollars, the commission rate drops to 1.6%. Only those orders not in compliance will be received by the subscriber:

```
String selector =
   "((Shares * Price) < 1000000.00 AND Commission > (Shares * Price) * .023) OR "
 + "((Shares * Price) >= 1000000.00 AND Commission > (Shares * Price) * .016)";

TopicSubscriber subscriber =
   session.createSubscriber(topic, selector, false);
```

Not Delivered Semantics

What happens to messages that are not selected for delivery to the consumer by its message selector? This depends on the message model used. For the publish-and-subscribe model, the messages are not delivered to that subscriber; they are, however, delivered to other pub/sub subscribers. This is true for both nondurable and durable subscriptions. For the point-to-point model, any messages that are not selected by the consumer are not visible to that consumer. They are, however, visible to other point-to-point consumers.

Because messages may not be delivered to consumers based on message filtering, it is important to make sure that all messages produced by a sender or publisher have a corresponding expiration associated with them. By default, messages are set never to expire. This means that if a message is filtered out and not delivered to a consumer, it will reside in the queue forever. By setting the message *time to live* option, you can control how long the message stays on the queue if it is not delivered.

The `JMSExpiration` header property contains a timestamp indicating when a message is set to expire. The JMS provider will constantly check the queue or topic (i.e., subscribers) for any messages set to expire and will automatically remove the message when the message is set to expire. The `JMSExpiration` header property is set by the JMS provider and is calculated by adding the time to live (in milliseconds) specified by the JMS developer to the timestamp when the message is sent or published.

The message's time to live property is set using the `MessageProducer` JMS interface, or more specifically, the `QueueSender` or `TopicPublisher` JMS interface. For example, to set all messages to expire 30 minutes after they have been sent and not delivered, you would set the message expiration as follows:

```
...
QueueSender sender = session.createSender(queue);
sender.setTimeToLive(1000 * 60 * 30);
...
```

Using the code snippet just shown, messages not delivered (due to message filters or consumers not available) would only remain on the queue or topic for 30 minutes. After that time, the JMS provider would remove the message from the queue or topic. While it is necessary to take into account the business requirements when deciding on the message expiration, it is equally important to make sure messages that are not delivered due to message filtering do not remain on the queue or topic for extended periods of time. Messages that are not delivered add to the queue depth and can cause unwanted queue depth notifications if left unattended for too long.

Design Considerations

There are two main message filtering approaches to consider when designing message-based solutions. You can send all related messages (e.g., trade orders) to a single JMS destination and use message filtering on the message consumer to select specific messages or you can use multiple JMS destinations that contain prefiltered messages. The first approach we will call the *Message Filtering approach*, and the second we will call the *Multiple Destination approach*. Understanding the implications of each of these approaches will help you arrive at a proper design for your particular situation.

What we have been focusing on so far in this chapter has been the Message Filtering approach using message filters on the `QueueReceiver` or `TopicSubscriber` to receive only those messages we are interested in. With the Message Filtering approach, messages are sent to a single JMS destination and filtered by the message consumer, as shown in Figure 6-1.

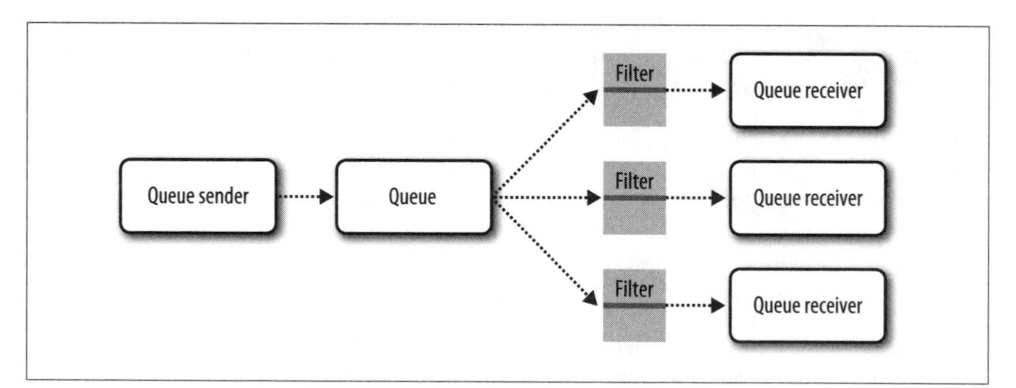

Figure 6-1. Message Filtering approach

Notice that with the Message Filtering approach it is the *message consumer* that has control over filtering and what messages it wants to receive. This approach provides for a higher level of decoupling between the message producer components and the message consumer components because less information needs to be known by the message producer about how the message will be processed. This is particularly true for the publish-and-subscribe model, where the topic publisher is generally unaware of the number and type of subscribers for a particular topic.

The Multiple Destination approach applies filtering *before* the message is sent to the destination. Rather than using message selectors, multiple destinations containing specific messages would be used instead. The message producer would typically use Java code to apply filtering logic to determine to which destination the message should be sent. Since each queue or topic contains a specific type of prefiltered message, the message consumer does not have to use message filtering to receive the messages it is interested in. This approach is illustrated in Figure 6-2.

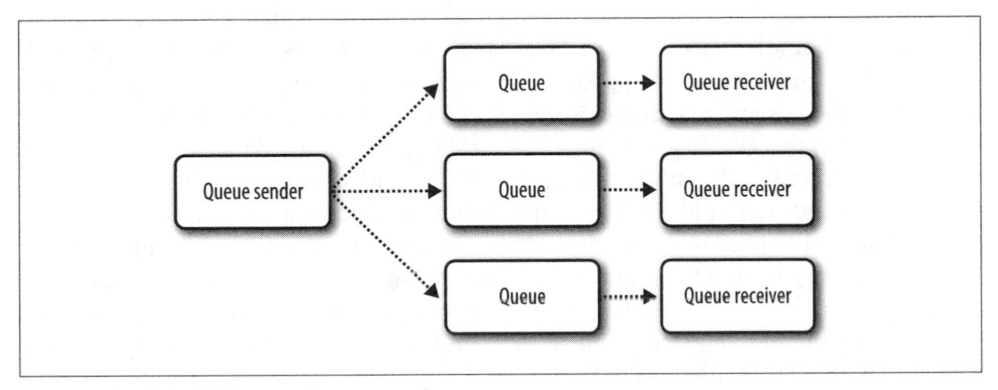

Figure 6-2. Multiple Destination approach

As shown in Figure 6-2, with the Multiple Destination approach it is the *message producer* that has control over the filtering and which destinations are to receive which messages. This is one of the fundamental differences between the two filtering approaches. One key factor to consider with the Multiple Destination approach is whether the message producer has enough knowledge of how the messages are processed to make the decision about which destinations to route the message to. The more the message producer needs to know about how the message will be consumed, the tighter the coupling will be between the message producer and message consumer. Of course, just because you are using the Multiple Destination approach does not mean that you cannot do further message selector-based filtering on the message consumer. The combined approach is illustrated in Figure 6-3.

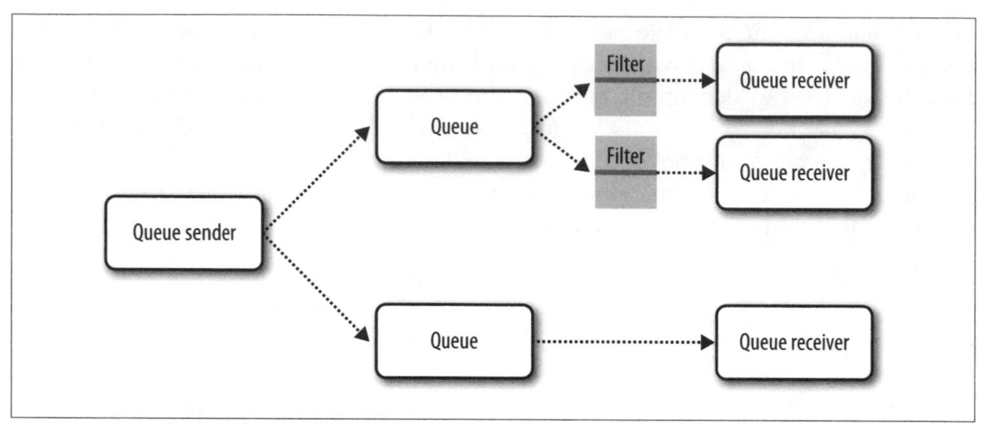

Figure 6-3. Combined approach

In many respects, the combined approach strikes the best balance between the two approaches. It resolves many of the issues facing both the Message Filtering approach and the Multiple Destination approach. Heavy use of either approach is a good indication that there is an issue in the overall design of your queues and topics, primarily with respect to the level of granularity of the JMS destinations. Topic subscribers containing a substantial amount of message filtering would suggest that the topics are too coarse-grained, and should perhaps be split up into multiple topics. On the other hand, topic subscribers forced to subscribe to multiple topics to retrieve the information they need would suggest that the topics are too fine-grained and should be combined to fit the majority of the use cases used by the topic subscribers. In general, the level of granularity represented by the queue or topic and overall queue or topic design should reflect how the information is used.

There are further implications within the point-to-point model with respect to message filtering, particularly when using the Message Filtering approach. With the point-to-point model, the message is guaranteed to be delivered to at most one message consumer. This means that if you are not careful with your message filtering on the message consumer side, there is the chance that the message selectors will be written so that the message will not be delivered to any of the consumers. For example, consider the case where message filtering is used to separate the processing of large and small stock trade orders. Two separate message consumers might have the corresponding message selectors:

```
//Consumer 1
String filter = "Shares < 5000";

//Consumer 2
String filter = "Shares > 5000";
```

If a trade order came through with exactly 5,000 shares, it would not be delivered to either consumer. The message would instead remain on the queue indefinitely, depending on the message expiry. If the message did expire, the trade order would be lost and never processed. That is bad news if the execution of that trade would have resulted in a substantial gain for the customer (but good news if it would have resulted in a loss!).

When using the Message Filtering approach with the point-to-point model, it is sometimes useful to create an *undelivered message consumer* so that messages excluded from the message filtering will be received by at least one message consumer. The message selector for the undelivered message consumer would be an exact negative of the union of all message selectors for all consumers of that queue. For example, if one queue receiver had a message selector specified as `CustType = 'GOLD'` and another had a message selector specified as `CustType = 'SILVER'`, the undelivered message consumer would specify its message selector as `CustType <> 'GOLD' AND CustType <> 'SILVER'`. This way, if a message came through without a `CustType` property or a `CustType` value other than `GOLD` or `SILVER`, the message would still be delivered. Upon receiving the message, the *undelivered message consumer* could then send out a notification to operations support, send the message to a workflow queue, or simply send the message back to the message producer with an associated error message.

To further illustrate the implications between the two filtering approaches, consider the example we have been using where an order is sent from a `Retailer` class and is processed by separate wholesaler consumers (in this case `WholesalerGold` and `WholesalerSilver`) based on the retailer type (`GOLD` or `SILVER`). Gold retailers go through a special order process that is significantly different from Silver retailers, hence the need for separate message consumers. With the Message Filtering approach, all messages are sent to a single order queue and delivered to the message consumers using message filtering by the `WholesalerGold` and `WholesalerSilver` classes.

First, the `Retailer` class sends an order, setting the message property based on the customer type of that retailer:

```
public class Retailer {
    ...
    private void makePurchase (String itemDesc, long qty, String custType) {
        try {
           ...

           Queue orderQ = (Queue)ctx.lookup("orderQueue");
           QueueSession session =
              connection.createQueueSession(false, Session.AUTO_ACKNOWLEDGE);
           QueueSender sender = session.createSender(orderQ);

           StreamMessage msg = session.createStreamMessage();
           msg.writeString(itemDesc);
           msg.writeLong(qty);

           msg.setStringProperty("CustType", custType);

           sender.send(msg);
        } catch (JMSException jmse) {
           ...
        }
```

```
        }
        ...
    }
```

Notice how the retailer only needs to be concerned about writing to one queue and how the `CustType` property is set on the message using an application property on the message. Next, the `WholesalerGold` class receives orders, but only for those orders having the `CustType` of `GOLD`:

```
public class WholesalerGold implements MessageListener {
   ...
   public WholesalerGold(String broker, String username, String password) {
      try {
         ...
         Queue orderQ = (Queue)ctx.lookup("orderQueue");

         QueueReceiver qReceiver =
             qSession.createReceiver(orderQ, "CustType = 'GOLD'");
         qReceiver.setMessageListener(this);

        } catch  (javax.jms.JMSException jmse) {
           ...
        }
   }
   ...
}
```

Finally, the `WholesalerSilver` class, which is listening on the same queue as the `WholesalerGold` class, receives orders only for retailers with a `CustType` of `SILVER`:

```
public class WholesalerSilver implements MessageListener {
   ...
   public WholesalerSolver(String broker, String username, String password) {
      try {
         ...
         Queue orderQ = (Queue)ctx.lookup("orderQueue");

         QueueReceiver qReceiver =
             qSession.createReceiver(orderQ, "CustType = 'SILVER'");
         qReceiver.setMessageListener(this);

         ...
      }
   }
   ...
}
```

Now let's take the same example but apply the Multiple Destination approach instead. In this case, the `retailer` class must apply conditional logic (in Java) to determine which queue to send the message to:

```
public class Retailer {
    ...
    private void makePurchase (String itemDesc, long qty, String custType) {
        try {
```

```
            ...
            Queue goldQueue = (Queue)ctx.lookup("GoldOrderQueue");
            Queue silverQueue = (Queue)ctx.lookup("SilverOrderQueue");
            QueueSession session =
               connection.createQueueSession(false, Session.AUTO_ACKNOWLEDGE);

            QueueSender goldSender = session.createSender(goldQueue);
            QueueSender SilverSender = session.createSender(silverQueue);

            StreamMessage msg = session.createStreamMessage();
            msg.writeString(itemDesc);
            msg.writeLong(qty);

            if (custType.equals("GOLD") {
               goldSender.send(msg);
            } else {
               silverSender.send(msg);
            }

         } catch  (JMSException jmse) {
            ...
         }
      }
      ...
   }
```

Notice the need for multiple senders, multiple queues, and conditional logic to route the message. The wholesaler classes, however, are simpler in that each wholesaler class receives all messages from the queue, with no message selectors needed. First the `WholesalerGold` receiver, which is listening on the order queue holding `Gold` orders:

```
public class WholesalerGold implements MessageListener {
   ...
   public WholesalerGold(String broker, String username, String password) {
      try  {
         ...
         Queue orderQ = (Queue)ctx.lookup("GoldOrderQueue");

         QueueReceiver qReceiver = qSession.createReceiver(orderQ);
         qReceiver.setMessageListener(this);

         ...
      }
   }
   ...
}
```

Next we have the `WholesalerSilver` class listening on the `SilverOrdersQueue`. Notice that no message selectors are required:

```
public class WholesalerSilver implements MessageListener {
   ...
   public WholesalerSilver(String broker, String username, String password) {
      try  {
```

```
            ...
            Queue orderQ = (Queue)ctx.lookup("SilverOrderQueue");

            QueueReceiver qReceiver = qSession.createReceiver(orderQ);
            qReceiver.setMessageListener(this);

            ...
        }
    }
    ...
}
```

If you look closely at the source code of the preceding Retailer and Wholesaler example you will see an immediate advantage to the Message Filtering approach. This approach is much more *extensible* than the Multiple Destination approach. If we were to add a third customer type to the mix with a CustType of PLATINUM, all that would need to change with the Message Filtering approach is to add another Wholesaler consumer class to consume orders from PLATINUM retailers. However, with the Multiple Destination approach, not only would we have to add another Wholesaler class to process PLATINUM retailers, but we would also have to add another queue to hold PLATINUM orders and modify the retailer class to route orders to the PLATINUM queue. Clearly, Message Filtering is a much better approach for this use case.

As you can see, while message selectors may seem like a fairly straightforward topic, there are many design-related implications to consider when using them, particularly when using the point-to-point model. Understanding these implications and the design trade-offs will help you make the right decisions for your messaging infrastructure.

CHAPTER 7

Guaranteed Messaging and Transactions

Guaranteed messaging is more than just a mechanism for handling disconnected consumers. It is a crucial part of the messaging paradigm and is the key to understanding the design of a distributed messaging system. This chapter examines *why* guaranteed messaging works, including message acknowledgment protocols that are part of guaranteed messaging and how to use client acknowledgments in applications. This chapter will also cover the design patterns of JMS that enable you to build guaranteed messaging into applications, and discuss failure scenarios, the rules that apply to recovery, and how to deal with recovery semantics in a JMS application.

Guaranteed Messaging

Before we discuss the parts of guaranteed messaging, we need to review and define some new terms. A JMS client application uses the JMS API. Each JMS vendor provides an implementation of the JMS API on the client, which we call the *client runtime*. In addition to the client runtime, the JMS vendor also provides some kind of message "server" that implements the routing and delivery of messages. The client runtime and the message server are collectively referred to as the *JMS provider*. Regardless of the architecture used by a JMS provider, the logical parts of a JMS system are the same. The number of processes and their location on the network is unimportant for this discussion. (In Chapter 10, you will see that some providers use a multicast architecture in which there is no central server.) The upcoming sections make use of diagrams that describe the logical pieces and do not necessarily reflect the process architecture of any particular JMS provider.

A *provider failure* refers to any failure condition that is outside of the domain of the application code. It could mean a hardware failure that occurs while the provider is entrusted with the processing of a message, an unexpected exception, the abnormal end of a process due to a software defect, or network failures.

There are three main parts to guaranteed messaging: message autonomy, store-and-forward, and the underlying message acknowledgment semantics. Each of these concepts is discussed in the following sections.

Message Autonomy

Messages are self-contained autonomous entities. This fact needs to be foremost in your mind when designing a distributed messaging application. A message may be sent and re-sent many times across multiple processes throughout its lifetime. Each JMS client along the way will consume the message, examine it, execute business logic, modify it, or create new messages in order to accomplish the task at hand.

In a sense, a JMS client has a contract with the rest of the system: when it receives a message, it does its part of the processing and may deliver the message (or new message) to another topic or queue. When a JMS client sends a message, it has done its job. The messaging server guarantees that any other interested parties will receive the messages. This contract between the sender and the message server is much like the contract between a JDBC client and a database. Once the data is delivered, it is considered "safe" and out of the hands of the client.

Store-and-Forward Messaging

When messages are marked *persistent*, it is the responsibility of the JMS provider to utilize a store-and-forward mechanism to fulfill its contract with the sender. The storage mechanism is used for persisting messages to disk (or some other reliable medium) in order to ensure that the message can be recovered in the event of a provider failure or a failure of the consuming client. The implementation of the storage mechanism is up to the JMS provider. The messages may be stored centrally (as is the case with centralized architectures) or locally, with each sending or receiving client (the solution used by decentralized architectures). While some vendors may still use a flat-file storage mechanism, most vendors use a database. Some may also use an intelligent combination of both. The forwarding mechanism is responsible for retrieving messages from storage and subsequently routing and delivering them.

Message Acknowledgments and Failure Conditions

JMS specifies a number of acknowledgment modes. These acknowledgments are a key part of guaranteed messaging. A message acknowledgment is part of the protocol that is established between the client runtime portion of the JMS provider and the server. Servers acknowledge the receipt of messages from JMS producers, and JMS consumers acknowledge the receipt of messages from servers. The acknowledgment protocol allows the JMS provider to monitor the progress of a message so that it knows whether the message was successfully produced and consumed. With this information, the JMS provider can manage the distribution of messages and guarantee their delivery.

Message Acknowledgments

The message acknowledgment protocol is the key to guaranteed messaging, and support for acknowledgment is required by the semantics of the JMS API. This section provides an in-depth explanation of how the acknowledgment protocol works and its role in guaranteed messaging.

We will begin by examining the `AUTO_ACKNOWLEDGE` mode. We will revisit this discussion later as it pertains to `CLIENT_ACKNOWLEDGE`, `DUPS_OK_ACKNOWLEDGE`, and JMS-transacted messages. An understanding of the basic concepts of `AUTO_ACKNOWLEDGE` will make it easy to grasp the fundamental concepts of the other modes.

The acknowledgment mode is set on a JMS provider when a `Session` is created:

```
tSession =
   tConnect.createTopicSession(false, Session.CLIENT_ACKNOWLEDGE);

qSession =
   qConnect.createQueueSession(false, Session.DUPS_OK_ACKNOWLEDGE);
```

AUTO_ACKNOWLEDGE

We'll look at the `AUTO_ACKNOWLEDGE` mode from the perspective of a message producer, the message server, and the message consumer.

The message producer's perspective

Under the covers, the `TopicPublisher.publish()` or `QueueSender.send()` methods are synchronous. These methods are responsible for sending the message and blocking until an acknowledgment is received from the message server. Once an acknowledgment has been received, the thread of execution resumes and the method returns; processing continues as normal. The underlying acknowledgment is not visible to the client programming model. If a failure condition occurs during this operation, an exception is thrown and the message is considered undelivered.

The message server's perspective

The acknowledgment sent to the producer (sender) from the server means that the server has received the message and has accepted responsibility for delivering it. From the JMS server's perspective, the acknowledgment sent to the producer is not tied directly to the delivery of the message. They are logically two separate steps.* For persistent messages, the server writes the message out to disk (the *store* part of store-and-forward), then acknowledges to the producer that the message was received (see Figure 7-1). For nonpersistent messages, this means the server may acknowledge the sender as soon as it has received the message and has the message in memory. If there

* In reality, these two operations may likely happen in parallel, but that depends on the vendor.

are no subscribers for the message's topic, the message may be discarded depending on the vendor.

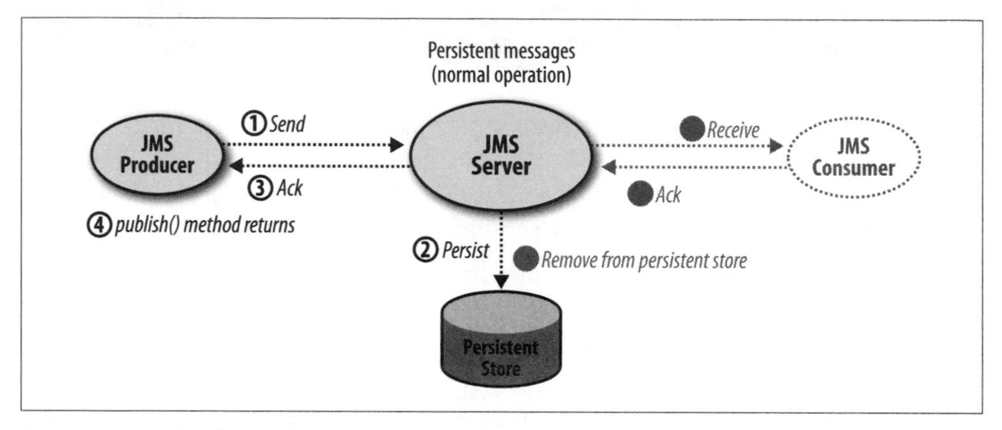

Figure 7-1. Send and receive are separate operations

In a publish-and-subscribe model, the message server delivers a copy of a message to each of the subscribers. For durable subscribers, the message server does not consider a message fully delivered until it receives an acknowledgment from all of the message's intended recipients. It knows on a per-consumer basis which clients have received each message and which have not.

Once the message server has delivered the message to all of its known subscribers and has received acknowledgments from each of them, the message is removed from its persistent store (see Figure 7-2).

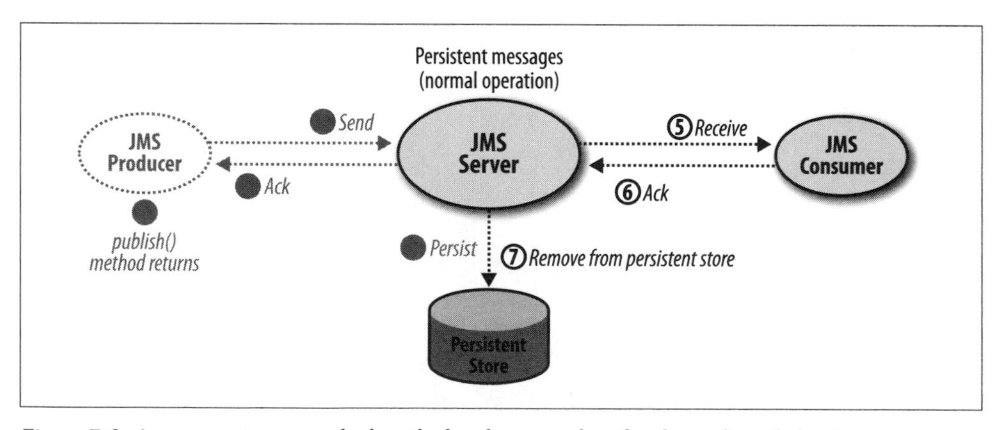

Figure 7-2. A message is removed when the last known subscriber has acknowledged

If the subscriptions are durable and the subscribers are not currently connected, then the message will be held by the message server until either the subscriber becomes available or the message expires. This is true even for nonpersistent messages. If a nonpersistent message is intended for a disconnected durable subscriber, the message server saves the message to disk as though it were a persistent message. In this case, the difference between persistent and nonpersistent messages is subtle, but very important. For nonpersistent messages, there may be a window of time after the message server has acknowledged the message to the sender and before it has had a chance to write the message out to disk on behalf of the disconnected durable subscribers. If the JMS provider fails during this window of time, the message may be lost (see Figure 7-3).†

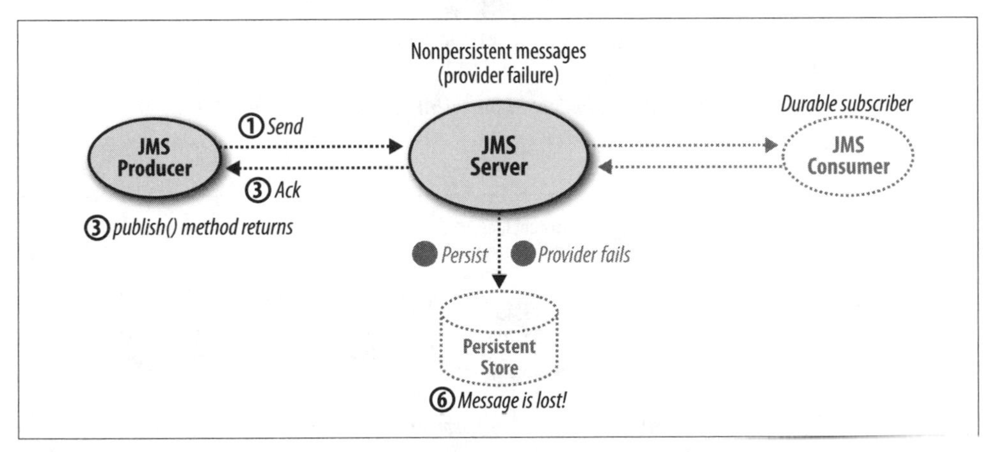

Figure 7-3. Nonpersistent messages with durable subscribers may be lost

With persistent messages, a provider may fail and recover gracefully, as illustrated in Figures 7-4 and 7-5. Since the messages are held in persistent storage, they are not lost, and will be delivered to consumers when the provider starts up again. If the messages are sent using a p2p queue, they are guaranteed to be delivered. If the messages were sent via publish-and-subscribe, they are guaranteed to be delivered only if the consumers' subscriptions are durable. The delivery behavior for nondurable subscribers may vary from vendor to vendor.

The message consumer's perspective

There are also rules governing acknowledgments and failure conditions from the consumer's perspective. If the session is in `AUTO_ACKNOWLEDGE` mode, the JMS provider's client runtime must automatically send an acknowledgment to the server as each consumer gets the message. If the server doesn't receive this acknowledgment, it considers the message undelivered and may attempt redelivery.

† In practice, the JMS provider may not allow this condition to happen. However, the JMS specification does imply that this failure condition can occur.

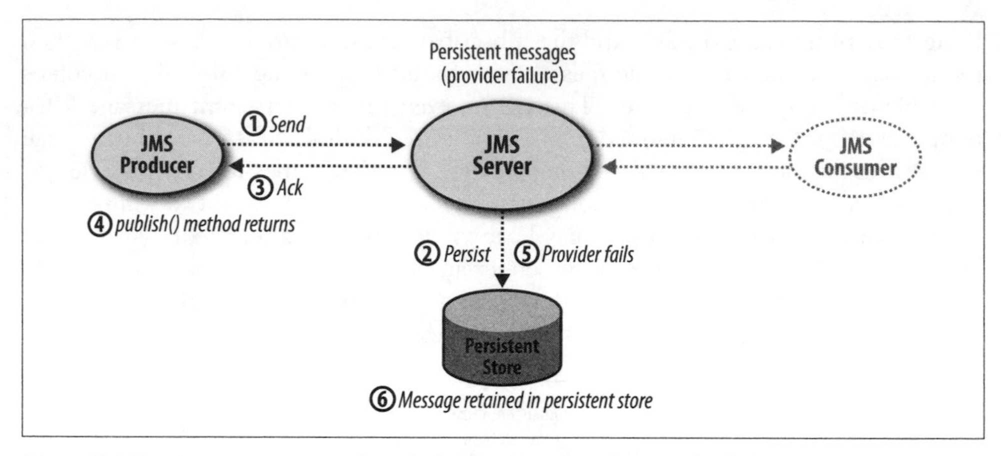

Figure 7-4. Persistent messages will not be lost in the event of a provider failure

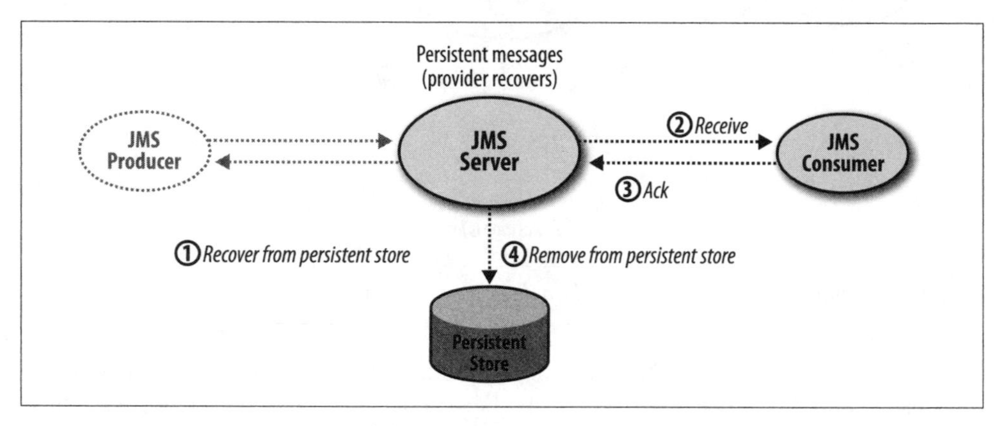

Figure 7-5. Persistent messages are delivered upon recovery of the provider

Message redelivery

The message may be lost if the provider fails while delivering a message to a consumer with a nondurable subscription. If a durable subscriber receives a message, and a failure occurs before the acknowledgment is returned to the provider (see Figure 7-6), then the JMS provider considers the message undelivered and will attempt to redeliver it (see Figure 7-7). In this case, the once-and-only-once requirement is in doubt. The consumer may receive the message again, because when delivery is guaranteed, it's better to risk delivering a message twice than to risk losing the message entirely. A redelivered message will have the `JMSRedelivered` flag set. A client application can check this flag by calling the `getJMSRedelivered` method on the `Message` object. Only the most recent message received is subject to this ambiguity.

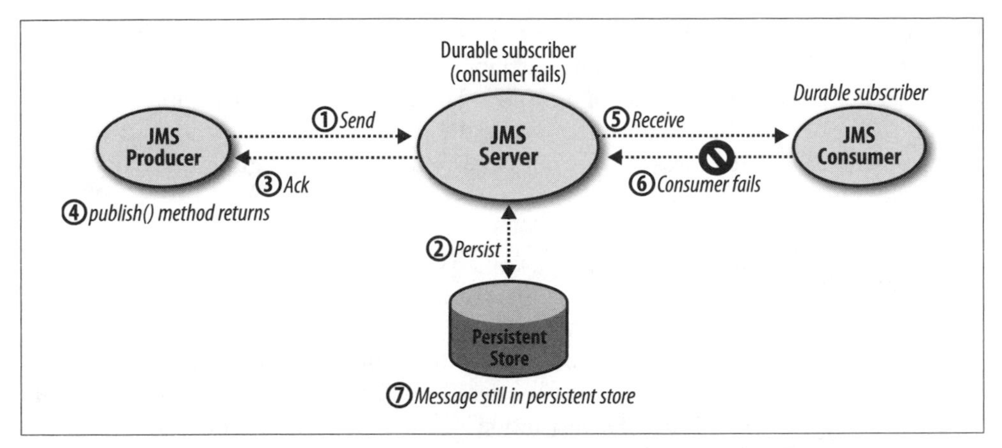

Figure 7-6. Failure occurs during the delivery of a message to a durable subscriber

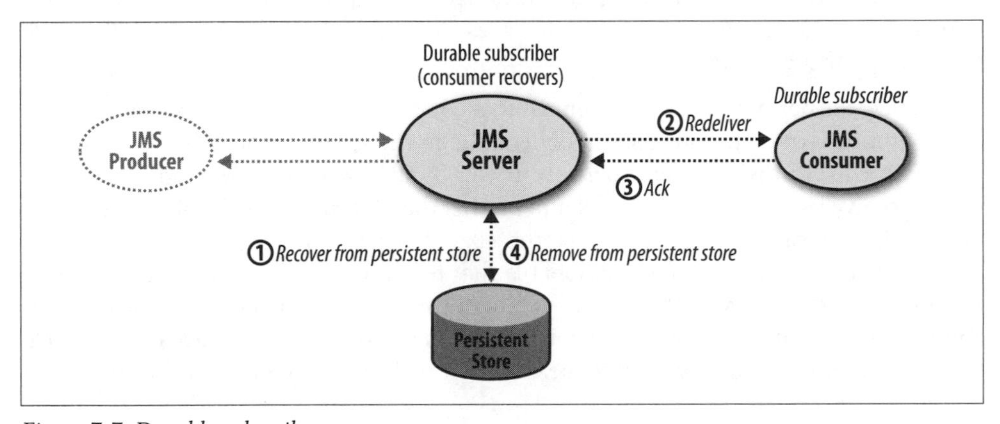

Figure 7-7. Durable subscriber recovers

To guard against duplicate messages while in `AUTO_ACKNOWLEDGE` mode, an application must check whether a redelivered message was already processed. One common technique for checking is to use a database table that is keyed on the `JMSMessageID` header. A `JMSMessageID` is unique for all messages and is intended for historical monitoring of messages in a repository. The `JMSMessageID` is therefore guaranteed to retain its uniqueness across provider failures. An alternate approach would be to use the `CLIENT_ACKNOWLEDGE` mode or to use a transacted message, which we will discuss in detail shortly.

Point-to-point queues

For point-to-point queues, messages are marked by the producer as either persistent or nonpersistent. If they are persistent, they are written to disk and subject to the same acknowledgment rules, failure conditions, and recovery as persistent messages in the publish-and-subscribe model.

From the receiver's perspective, the rules are somewhat simpler, since only one consumer can receive a particular instance of a message. A message stays in a queue until it is either delivered to a consumer or it expires. This is analogous to a durable subscriber in that a receiver can be disconnected while the message is being produced without losing the message. If the messages are nonpersistent, they are not guaranteed to survive a provider failure.

DUPS_OK_ACKNOWLEDGE

Specifying the `DUPS_OK_ACKNOWLEDGE` mode on a session instructs the JMS provider that it is OK to send a message more than once to the same destination. This is different from the once-and-only-once or the at-most-once delivery semantics of `AUTO_ACKNOWLEDGE`. The `DUPS_OK_ACKNOWLEDGE` delivery mode is based on the assumption that the processing necessary to ensure once-and-only-once delivery incurs extra overhead and hinders performance and throughput of messages at the provider level. An application that is tolerant of receiving duplicate messages can use the `DUPS_OK_ACKNOWLEDGE` mode to avoid incurring this overhead.

In practice, the performance improvement that you gain from `DUPS_OK_ACKNOWLEDGE` may be insignificant or even nonexistent, depending on the JMS vendor. It is even conceivable that a JMS provider could perform better in `AUTO_ACKNOWLDEGE` mode because it would receive its acknowledgments sooner rather than later. This could allow it to clean up resources more quickly, or reduce the size of persistent storage and in-memory queues. At first glance, it seems reasonable that fewer acknowledgments result in less network traffic. However, the network may not be the bottleneck under heavy load conditions with large numbers of clients. In summary, the benefits of `DUPS_OK_ACKNOWLEDGE` are something you may want to measure before designing your application around it.

CLIENT_ACKNOWLEDGE

With `AUTO_ACKNOWLEDGE` mode, the acknowledgment is always the last thing to happen implicitly after the `onMessage()` handler returns. The client receiving the messages can get finer-grained control over the delivery of guaranteed messages by specifying the `CLIENT_ACKNOWLEDGE` mode on the consuming session.

The use of `CLIENT_ACKNOWLEDGE` allows the application to control when the acknowledgment is sent. For example, an application can acknowledge a message—thereby relieving the JMS provider of its duty—and perform further processing of the data represented by the message. The key to this is the `acknowledge()` method on the `Message` object, as shown in the following example:

```
public void onMessage(javax.jms.Message message) {
    try {
        // Perform some business logic with the message
        ...
```

```
            message.acknowledge();
            // Perform more business logic with the message
            ...
        } catch (javax.jms.JMSException jmse) {
            // Catch the exception thrown and undo the results
            // of partial processing
            ...
        }
    }
```

The acknowledge() method informs the JMS provider that the message has been successfully received by the consumer. This method throws an exception to the client if a provider failure occurs during the acknowledgment process. The provider failure results in the message being retained by the JMS server for redelivery. Therefore, the exception handling code should undo the results of any partially processed business logic in preparation for receiving the message again, or it should log the message as processed so that the redelivered message can be ignored. The acknowledge() method should only be used with the CLIENT_ACKNOWLEDGE mode; if used with the AUTO_ACKNOWLEDGE or DUPS_OK_ACKNOWLEDGE mode, the call is ignored by the JMS provider.

Grouping multiple messages

The CLIENT_ACKNOWLEDGE mode also gives you the ability to batch together multiple message receipts and consume them in an all-or-nothing fashion. A consuming client may receive several messages in a sequence and treat them as a group. CLIENT_ACKNOWLEDGE does not provide the capability to consume messages selectively. A single acknowledgment for the last message in the group implicitly acknowledges all previously unacknowledged messages for the current session. This means that if the client application fails before the last message is acknowledged, it may recover when it comes back up. All of the unacknowledged messages will be resent with the JMSRedelivered flag set on each of the unacknowledged messages. A JMS client may also call the recover() method on a Session object to force the redelivery of all previously unacknowledged messages, even if there hasn't been a failure.

Message Groups and Acknowledgment

When multiple messages need to be dealt with as a group, the application needs to be able to store or cache interim messages until the entire group has been delivered. This requirement typically means that the asynchronous invocation of the onMessage() handler would result in business logic getting executed and data would be placed temporarily in a database table in preparation for processing the group of messages as a whole. When the last message of the group arrives, the application can then go to the database to retrieve the data from the previous messages to establish any context it may need.

Handling Redelivery of Messages in an Application

JMS provides strict rules that govern when the redelivered flag is set. In AUTO_ACKNOWLEDGE mode, only the most recently consumed message is subject to ambiguous redelivery. In other modes, multiple messages may have the redelivered flag set. It is up to the application designer to isolate the conditions under which ambiguity can occur and to account for it in the application. To illustrate the use of the CLIENT_ACKNOWLEDGE mode and redelivered flag, we will use a simple example where multiple messages are sent as a group to the receiver.

Whenever a message is redelivered to a message consumer, the JMSRedelivered message header property will be set to true on the message. This is an indication to the message consumer that the message has been partially processed but an exception occurred preventing that message from being completely processed. In some cases, this flag only serves to inform the message consumer that the message is being sent again, and no action needs to be taken by the message consumer. However, in other cases, specific action might need to be taken if the message is being redelivered to the message consumer.

In the example we are about to show, multiple messages may be sent to a message consumer as a group, meaning that the messages must be processed together and thus acknowledged by the message consumer as a group. This requires the use of the CLIENT_ACKNOWLEDGE mode so that the message consumer has control over when the messages are acknowledged and hence marked as having been delivered by the JMS provider. If a failure occurs during the processing of the message group, the entire message group is redelivered.

To check if a message has been redelivered, you can interrogate the JMSRedelivered message header property of the message received:

```
...
if (message.getJMSRedelivered()) {
   processCompensatingTransaction();
}
...
```

As you will see in the example provided in the next section, the contents of the message need to be stored in some sort of in-memory cache or persistence store when processing multiple messages in a group. If the message is being redelivered, the data in the persistence store or in-memory cache will need to be cleared because the message is being processed again. Without checking to see whether the message has been redelivered, you could end up with duplicate data in the cache or persistence store.

Message Groups Example

While there are several use cases for using the CLIENT_ACKNOWLEDGE mode and redelivery flag, one of the best examples is that of processing groups of messages. In this case, multiple messages are sent from the message producer, but need to be "batched" and processed by the message consumer as a single unit of work. A transacted session would not work in this case because the messages may be sent to the message consumer individually and at different times.

When processing messages as a group, typically the message producer will use some sort of custom sequence marker to indicate when the sequence of messages start and end. The ending sequence marker tells the message consumer that all of the messages in the group have been sent and it is OK to process the messages and then acknowledge that they have all been delivered. One way to do this is with an empty payload message containing a custom sequence marker message property:

```
...
//send an empty payload message starting the group
BytesMessage msg = session.createBytesMessage();
msg.setStringProperty("SequenceMarker", "START_SEQUENCE");
sender.send(msg);

//now send the messages
TextMessage msg = session.createTextMessage(messagePayload);
sender.send(msg);
...

//send an empty payload message ending the group
BytesMessage msg = session.createBytesMessage();
msg.setStringProperty("SequenceMarker", "END_SEQUENCE");
sender.send(msg);
```

The message consumer would start accumulating the payload data until it receives the `END_SEQUENCE` event message, at which point it would then process the messages. Once processing is complete, the message consumer would then acknowledge all of the messages using the `acknowledge()` method on the `Message` object.

To illustrate how to process groups of messages using the `CLIENT_ACKNOWLEGE` mode and the `JMSRedelivered` message header property, we will use two simple classes: `JMSSender` and `JMSReceiver`. The `JMSSender` class will connect to the JMS provider and send a group of three simple `TextMessages`, along with an empty `BytesMessage` containing a `START_SEQUENCE` message property and an empty `BytesMessage` containing an `END_SEQUENCE` message property.

The `JMSReceiver` class will then recognize when a sequence of messages is starting, and until the `END_SEQUENCE` message is received, will accumulate the message payload in a simple `List`. Once all of the messages have been received, the `JMSReceiver` class will then display the message payload for all of the messages and acknowledge the receipt of the messages.

We will start by showing the complete listing for the `JMSSender` and `JMSReceiver` classes and then describe the areas of the code that pertain to the `CLIENT_ACKNOWLEDGE` mode and redelivery flag. The following is the complete listing of the `JMSSender` class:

```
import javax.jms.*;
import javax.naming.*;
```

```
public class JMSSender {

  private QueueConnection connection = null;
  private QueueSession session = null;
  private QueueSender sender = null;

  public static void main(String[] args) {
    try {
      JMSSender app = new JMSSender();
      app.sendMessageGroup();
      System.exit(0);
    } catch (Exception ex) {
        ex.printStackTrace();
    }
}

  public JMSSender() {
    try {
      //connect to the jms provider and create the
      //connection, session, and sender
      Context ctx = new InitialContext();
      QueueConnectionFactory factory = (QueueConnectionFactory)
          ctx.lookup("QueueCF");
      connection = factory.createQueueConnection();
      connection.start();
      session =
        connection.createQueueSession(false, Session.AUTO_ACKNOWLEDGE);
      Queue queue = (Queue)ctx.lookup("queue1");
      sender = session.createSender(queue);
    } catch (Exception jmse) {
       jmse.printStackTrace();
    }
  }

  public void sendMessageGroup() throws JMSException {
    //send the messages as a group
    sendSequenceMarker("START_SEQUENCE");
    sendMessage("First Message");
    sendMessage("Second Message");
    sendMessage("Third Message");
    sendSequenceMarker("END_SEQUENCE");
    connection.close();
}

  //send a simple text message within the group of messages
  private void sendMessage(String text) throws JMSException {
    TextMessage msg = session.createTextMessage(text);
    msg.setStringProperty("JMSXGroupID", "GROUP1");
    sender.send(msg);
  }

  //send an empty payload message containing the sequence marker
  private void sendSequenceMarker(String marker) throws JMSException {
    BytesMessage msg = session.createBytesMessage();
```

```
        msg.setStringProperty("SequenceMarker", marker);
        msg.setStringProperty("JMSXGroupID", "GROUP1");
        sender.send(msg);
      }
    }
```

The main method of the JMSSender class instantiates a new JMSSender object and invokes the main driver method sendMessageGroup. The constructor establishes a connection to the JMS provider and creates the QueueConnection, QueueSession, Queue, and QueueSender objects needed by the rest of the class. The sendMessageGroup method sends a blank message containing the starting sequence marker, then sends three simple messages, then the ending sequence marker. It then closes the connection. The sendMessage method creates a simple TextMessage and sends it to the queue, whereas the sendSequenceMarker method sends an empty BytesMessage containing the sequence marker. We will describe this code in more detail later in this section.

The complete listing for the JMSReceiver class is given here:

```
public class JMSReceiver implements MessageListener {

  private List<String> messageBuffer = new ArrayList<String>();

  public JMSReceiver() {
    try {
      Context ctx = new InitialContext();
      QueueConnectionFactory factory = (QueueConnectionFactory)
         ctx.lookup("QueueCF");
      QueueConnection connection = factory.createQueueConnection();
      connection.start();
      QueueSession session =
         connection.createQueueSession(false, Session.CLIENT_ACKNOWLEDGE);
      Queue queue = (Queue)ctx.lookup("queue1");
      QueueReceiver receiver = session.createReceiver(queue);
      receiver.setMessageListener(this);
    } catch (Exception ex) {
       ex.printStackTrace();
    }
  }

  public void onMessage(Message message) {
    try {
      if (message.propertyExists("SequenceMarker")) {
        String marker = message.getStringProperty("SequenceMarker");

        //if we are starting a message group, clear out the message buffer
        if (marker.equals("START_SEQUENCE")) {
          //since the messages are delivered and acknowledged as a group, any
          //failures will result in the first sequence message being marked as
          //being redelivered - we don't care about the others
          if (message.getJMSRedelivered()) {
            processCompensatingTransaction();
          }

          messageBuffer.clear();
```

```
        }

        //if we are ending the message group, process the message and
        //acknowledge that all messages as having been delivered
        if (marker.equals("END_SEQUENCE")) {
          //process the message
          System.out.println("Messages: ");
          for (String msg : messageBuffer) {
            System.out.println(msg);
          }

          //acknowledge that all messages have been received
          message.acknowledge();
        }
      }

      //save the message contents if it is a non-marker message
      if (message instanceof TextMessage) {
        TextMessage msg = (TextMessage)message;
        processInterimMessage(msg.getText());
      }

      //wait for the next message
      System.out.println("waiting for messages...");
    } catch (Exception ex) {
       ex.printStackTrace();
       System.exit(1);
    }
  }

  public void processCompensatingTransaction() {
    //reverse the processing from the prior message set
    messageBuffer.clear();
  }

  public void processInterimMessage(String msg) {
    //process the interim message
    messageBuffer.add(msg);
  }

  public static void main(String argv[]) {
     new JMSReceiver();
  }
}
```

In the preceding `JMSReceiver` class, the constructor establishes a connection to the JMS provider and creates the `QueueConnection`, `QueueSession`, `Queue`, and `QueueReceiver` objects needed by the rest of the class. The bulk of the work occurs in the `onMessage` method, where messages are accumulated and processed.

The processing of the message group starts with the `JMSSender` class sending the various messages within the message group:

```
public void sendMessageGroup() throws JMSException {
  //send the messages as a group
```

```
    sendSequenceMarker("START_SEQUENCE");
    sendMessage("First Message");
    sendMessage("Second Message");
    sendMessage("Third Message");
    sendSequenceMarker("END_SEQUENCE");
    connection.close();
}
```

The `JMSSender` first sends an empty `BytesMessage` containing a sequence marker value of `START_SEQUENCE`. Note that although we chose to use a `BytesMessage`, any JMS message type could have been used to send the sequence marker message. Then the three `TextMessage` messages are sent, followed by another empty `BytesMessage` containing the `END_SEQUENCE` sequence marker. Note that all five of these messages are sent as separate messages to the JMS provider (and in turn are received as separate messages in the `JMSReceiver`).

Once the `START_SEQUENCE` message is sent, it is picked up by the `JMSReceiver` and processing of the message group begins:

```
public void onMessage(Message message) {
  try {
    if (message.propertyExists("SequenceMarker")) {
      String marker = message.getStringProperty("SequenceMarker");

      ...
      //if we are starting a message group, clear out the message buffer
      if (marker.equals("START_SEQUENCE")) {
        ...
        messageBuffer.clear();
      }
  ...
}
```

The `onMessage` method of the `JMSReceiver` class first detects that the `SequenceMarker` property exists on the message, indicating that it is an event message containing a sequence directive. The `START_SEQUENCE` message causes the `messageBuffer` `ArrayList` to be cleared, preparing it to accept the payload from the message group.

The `JMSSender` class then sends the three `TextMessage` objects, which are picked up by the `JMSReceiver`. The `JMSReceiver` recognizes these `TextMessage` objects as being part of the message group, and invokes the `processInterimMessage` method, which adds the message payload to the `messageBuffer` `ArrayList` object:

```
public void onMessage(Message message) {
  try {
    //save the message contents if it is a non-marker message
    if (message instanceof TextMessage) {
      TextMessage msg = (TextMessage)message;
      processInterimMessage(msg.getText());
    }
  ...
}
```

```
public void processInterimMessage(String msg) {
  //process the interim message
  messageBuffer.add(msg);
}
```

In real life, one would imagine the processInterimMessage method inserting the message payload into a staging database or a sophisticated in-memory cache. However, for purposes of this example, we are simply adding the message contents to an ArrayList represented by the messageBuffer attribute.

The last message sent by the JMSSender class is the END_SEQUENCE message, indicating that the sequence of messages is over and can be processed. The JMSReceiver class picks up this message and, upon receipt of the message, prints out the contents of the messages and acknowledges that the messages have been delivered:

```
public void onMessage(Message message) {
  try {
    if (message.propertyExists("SequenceMarker")) {
       String marker = message.getStringProperty("SequenceMarker");
       ...

       //if we are ending the message group, process the message and
       //acknowledge all messages as having been delivered
       if (marker.equals("END_SEQUENCE")) {
         //process the message
         System.out.println("Messages: ");
         for (String msg : messageBuffer) {
           System.out.println(msg);
         }

       //acknowledge that all messages have been received
       message.acknowledge();
    }
  }
  ...
}
```

Calling the acknowledge method on a message acknowledges the current message *and all previously unacknowledged messages*. Because the logic in the JMSReceiver class does not process the messages until it sees the END_SEQUENCE sequence marker message, it also does not explicitly acknowledge the receipt of the message until it knows it can process all the messages at the same time. This logic avoids processing the first message if the second message fails to be delivered, and so on. If the messages were to be separately acknowledged, the client could fail after the first message was acknowledged, but before the second or third message was fully processed. If this occurred, the first message would be considered delivered by the JMS provider, yet not fully processed by the client. It would be effectively lost. Delaying acknowledgment provides a way to write the application so that it behaves correctly when failures occur.

A single acknowledgment is now sent, acknowledging all three messages. The JMS provider has now fulfilled its part of the contract with the receiving application and can remove the messages from its persistent store.

Running these two classes produces the following output:

```
$ ./jmsreceiver.sh
Messages:
First Message
Second Message
Third Message
waiting for messages...
```

So, what happens when an exception occurs during the processing of this message group or the message consumer failed to invoke the `acknowledge` method after the last `END_SEQUENCE` message? In both these cases, the messages that were previously delivered to the message consumer would be marked as not having been delivered and the delivery of all of the messages in the group (including the sequence marker messages) would be redelivered to the message consumer. In this case, the messages that were previously delivered to the message consumer would have the `JMSRedelivered` message header property set to `true` (it has a value of `false` otherwise).

Let's modify the message processing in the `JMSReceiver` class to throw an exception when the third message in the group is received:

```
public void onMessage(Message message) {
  try {
    //save the message contents if it is a non-marker message
    if (message instanceof TextMessage) {
      TextMessage msg = (TextMessage)message;
      processInterimMessage(msg.getText());
      if (msg.getText().equals("Third Message")) {
        throw new Exception("Exception after Message 3");
      }
    }

    //wait for the next message
    System.out.println("waiting for messages...");
  } catch (Exception ex) {
     ex.printStackTrace();
     System.exit(1);
  }
}
```

Running the two classes with the code change just shown would produce the following output:

```
$ ./jmsreceiver.sh
Messages:
First Message
Second Message
java.lang.Exception: Exception after Message 3
```

If we browse the queue1 queue, we would find the following messages waiting to be redelivered:

```
$ ./jmsbrowser.sh
------------------------------------------
Message: START_SEQUENCE, Redelivered: true
------------------------------------------
Message: First Message, Redelivered: true
------------------------------------------
Message: Second Message, Redelivered: true
------------------------------------------
Message: Third Message, Redelivered: true
------------------------------------------
Message: END_SEQUENCE, Redelivered: false
```

Notice that all the messages except the END_SEQUENCE message have the JMSRedelivered header property set to true. Since the last message was never received by the message consumer, it was not marked as having been redelivered.

We now have an issue: Since we are storing the message contents for this message group, if the JMSReceiver class were to pick up these messages a second time, we could, in real life, get duplicate data processed by the message group. The JMSReceiver class therefore needs to take corrective action in the event the messages are redelivered:

```
public void onMessage(Message message) {
  try {
    if (message.propertyExists("SequenceMarker")) {
      String marker = message.getStringProperty("SequenceMarker");

      //if we are starting a message group, clear out the message buffer
      if (marker.equals("START_SEQUENCE")) {
        if (message.getJMSRedelivered()) {
          processCompensatingTransaction();
        }

        messageBuffer.clear();
      }
   ...
}

public void processCompensatingTransaction() {
  //reverse the processing from the prior message set
  messageBuffer.clear();
}
```

Notice that the onMessage method of the JMSReceiver first checks to see whether the message was being redelivered by invoking the getJMSRedelivered method on the message object. Since the messages are delivered and acknowledged as a group, any failures in the onMessage method will result in the entire message sequence being redelivered. As a result, we only have to check the redelivered flag for the first message. Checking the redelivered flag for subsequent messages would result in the message buffer continually being reloaded.

In the preceding code, if the redelivered flag was set on the message, we invoked the processCompensatingTransaction method to reverse any persisted or cached changes as a result of the prior message processing. In our case, the call to processInterimMessage really doesn't do anything except add the message payload to an internal ArrayList. However, in a real-world application, the call to processInterimMessage would probably execute some business logic and place data in a database table in preparation for the next message. Upon failure of the client, the messages would be redelivered, and the processCompensatingTransaction would clean up or reinitialize any application-specific data that may have been left in an unclean state.

This is a good argument for message autonomy. Each message should be self-contained. When multiple messages need to depend on each other, the application should be written like a finite state machine where the results of the processing of one message are saved so that the application's state can be reestablished at a later time. The next message can then independently reestablish all of the context it needs to do its work. This is a perfectly viable and valid application design and should be considered in lieu of, or in conjunction with, other approaches.

Message Grouping and Multiple Receivers

There are additional factors to take into account for message grouping when using multiple receivers. As you learned in Chapter 4, the point-to-point messaging model supports load balancing through the use of concurrent message listeners (i.e., multiple receivers), allowing the messaging system to process multiple messages at the same time. As you can guess, this poses a serious problem when grouping messages. Without the proper code and message properties set, you run the risk of having the group of messages spread across multiple receivers. This would cause obvious problems in that one receiver may get the first couple of messages, whereas another receiver ends up getting the message containing the group terminator. This would result in messagages being left on the queue without ever being acknowledged. Fortunately, there is a simple solution to this problem.

Whenever you use message grouping, you should always use the JMSXGroupID message property. This property will essentially provide you with a "sticky consumer," guaranteeing that all messages in the group go to the same consumer, regardless of the number of consumers listening on the queue. Notice in the code example from the previous section that the sendMessage() and the sendSequenceMarker() methods of the JMSSender class set an additional JMSXGroupID property on the message containing the value GROUP1. This value is completely arbitrary and could be assigned any value whatsoever:

```
import javax.jms.*;
import javax.naming.*;

public class JMSSender {
```

```
  ...

  //send a simple text message within the group of messages
  private void sendMessage(String text) throws JMSException {
    TextMessage msg = session.createTextMessage(text);
    msg.setStringProperty("JMSXGroupID", "GROUP1");
    sender.send(msg);
  }

  //send an empty payload message containing the sequence marker
  private void sendSequenceMarker(String marker) throws JMSException {
    BytesMessage msg = session.createBytesMessage();
    msg.setStringProperty("SequenceMarker", marker);
    msg.setStringProperty("JMSXGroupID", "GROUP1");
    sender.send(msg);
  }
}
```

When the `JMSXGroupID` property is set, the JMS provider will look for a consumer that has that group ID assigned to it. If there are no consumers assigned to that group, the JMS provider will pick one based on its load balancing scheme and assign it the group ID. From that point on, only that consumer will receive the messages associated with that group.

Notice that this property is a JMS extension (hence the `JMSX` prefix). As such, there is no corresponding `setJMSXGroupID()` method as with the other JMS header properties. There's a corresponding JMS extension property that is related to the `JMSXGroupID` called the `JMSXGroupSeq` message property. This `int` property is an optional property that you can use to specify a sequence within the group. Referring to the example in the previous section, in the `sendMessage()` method of the `JMSSender` class we could have specified a sequence of messages using the `JMSXGroupSeq` property as follows:

```
public class JMSSender {

  ...

  //send a simple text message within the group of messages
  private void sendMessage(String text, int sequence) throws JMSException {
    TextMessage msg = session.createTextMessage(text);
    msg.setIntProperty("JMSXGroupSeq", sequence);
    msg.setStringProperty("JMSXGroupID", "GROUP1")
    sender.send(msg);
  }

  ...
}
```

Since the messages would be sent to the queue as an ordered list and only received by a single "sticky" consumer, this property is usually skipped. However, it is a useful property in that it is one way to essentially "close" the group. If you are done processing the group and want to free up the consumer for another group, you can close the group by setting the `JMSXGroupID` to a value of –1.

Transacted Messages

Our discussion of message acknowledgment shows that producers and consumers have different perspectives on the messages they exchange. The producer has a contract with the message server that ensures the message will be delivered as far as the server. The server has a contract with the consumer that ensures the message will be delivered to it. The two operations are separate, which is a key benefit of asynchronous messaging. It is the role of the JMS provider to ensure that messages get to where they are supposed to go. Having all producers and all consumers participate in one global transaction would defeat the purpose of using a loosely coupled asynchronous messaging environment.

JMS transactions follow the convention of separating the send operations from the receive operations. Figure 7-8 shows a transactional send, in which a group of messages are guaranteed to get to the message server, or none of them will. From the sender's perspective, the messages are cached by the JMS provider until a `commit()` is issued. If a failure occurs, or a `rollback()` is issued, the messages are discarded. Messages delivered to the message server in a transaction are not forwarded to the consumers until the producer commits the transaction.

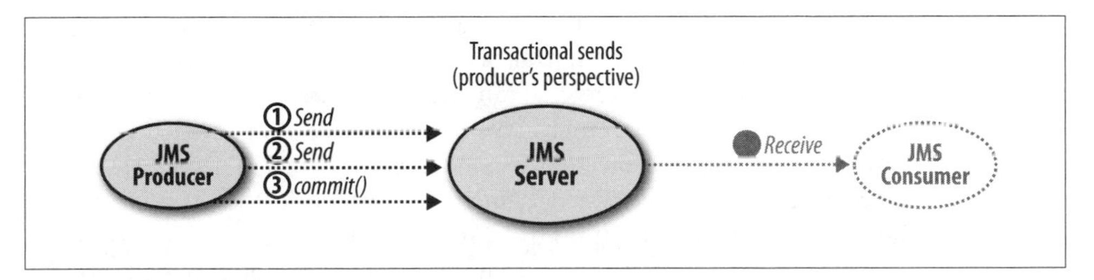

Figure 7-8. Transactional messages are sent in an all-or-nothing fashion

The JMS provider will not start delivery of the messages to its consumers until the producer has issued a `commit()` on the session, even though it has received all of the messages from the sender. The scope of a JMS transaction can include any number of messages. Although similar in concept, the session `commit()` is not the same as a Java Transaction API (JTA) transaction `commit()`. The session transaction is managed by the JMS provider, not JTA.

It should be no surprise that JMS also supports transactional receives, in which a group of transacted messages are received by the consumer on an all-or-nothing basis (see Figure 7-9). From the transacted receiver's perspective, the messages are delivered to it as expeditiously as possible, yet they are held by the JMS provider until the receiver issues a `commit()` on the session object. If a failure occurs or a `rollback()` is issued, then the provider will attempt to redeliver the messages, in which case the messages will have the redelivered flag set.

When sending messages, if the `Session.commit()` method is *not* invoked upon normal completion of the method sending the messages, the messages are removed from the queue by the JMS provider and never delivered to the message consumers. When receiving messages, if the `Session.commit()` method is not invoked upon normal completion of the method receiving the messages, the messages are marked as having not been delivered. The JMS provider will redeliver the messages to the message consumer with the `JMSRedelivered` header property set to `true`, indicating that there was a prior attempt to process the messages. Thus, in either case, care must be taken to ensure that the session is either committed or rolled back prior to the end of the method.

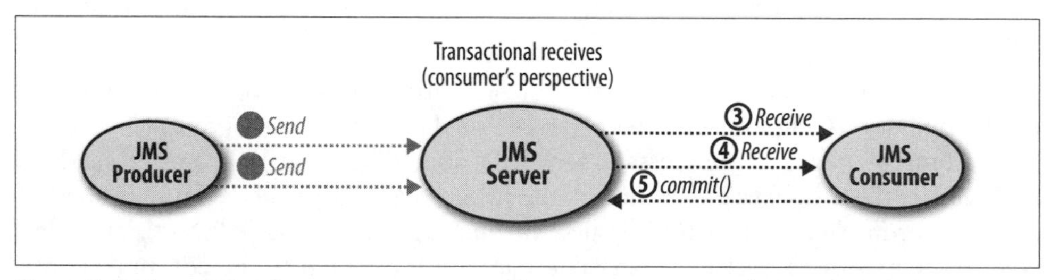

Figure 7-9. Transactional messages are received by a consumer in an all-or-nothing fashion

Transacted producers and consumers can exchange messages with nontransacted consumers and producers. The scope of the transaction is limited to the producer's or consumer's session with the message server. Transacted producers and transacted consumers can, however, be grouped together in a single transaction, provided that they are created from the same session object, as shown in Figure 7-10. This allows a JMS client to produce and consume messages as a single unit of work. If the transaction is rolled back, the messages produced within the transaction will not be delivered by the JMS provider. The messages consumed within the same transaction will not be acknowledged and will be redelivered.

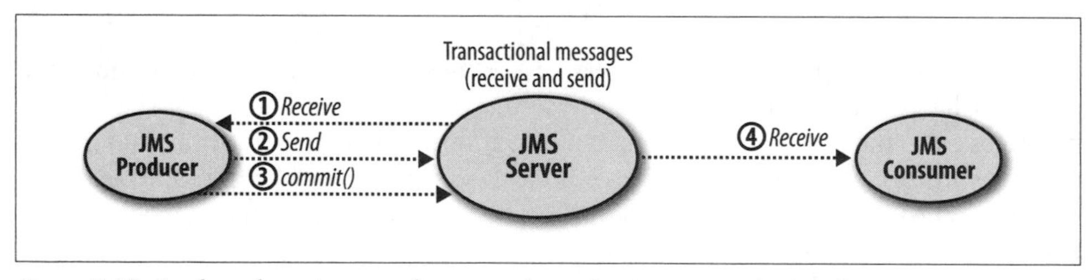

Figure 7-10. Sends and receives may be grouped together in one transactional session

Unless you are doing a synchronous request/reply, you should avoid grouping a send followed by an asynchronous receive within a transaction. In general, transactions should be as short-lived as possible. When using message grouping, there could be a long interval between the time that a message is sent and the related message is asynchronously received, depending on failures or downtime of other processes that are involved. It is more practical to group the receipt of a message with the send of another message.

Creating and Using a JMS Transaction

Now that you understand the concepts of transactional sends and receives, we can take a look at some code. The first step in creating a transactional message is the initialization of the `Session` object:

```
// Create a transacted TopicSession
TopicSession session =
   connect.createTopicSession(true, Session.AUTO_ACKNOWLEDGE);

// Create a transacted QueueSession
QueueSession =
   connect.createQueueSession(true, Session.AUTO_ACKNOWLEDGE);
```

The first parameter of a `createTopicSession()` or `createQueueSession()` method is a `boolean` indicating whether this is a transacted session. That is the only thing required to create a transactional session. There is no explicit `begin()` method. When a session is transacted, all messages sent or received using that session are automatically grouped in a transaction. The transaction remains open until either a `session.rollback()` or a `session.commit()` happens, at which point a new transaction is started.‡ An additional `Session` method, `getTransacted()`, returns `true` or `false` indicating whether or not the current session is transactional.

Transacted Session Example

In this very simple example, we will demonstrate the use of the transacted session for sending a group of messages. When sending messages in a transacted session, the messages will not be delivered to the message consumer until the session in the message producer has been committed via the `Session.commit()` method:

```
import javax.jms.*;
import javax.naming.*;

public class JMSSenderTransacted {

  private QueueConnection connection = null;
  private QueueSession session = null;
  private QueueSender sender = null;

  public void sendMessages() {
    try {
      //send the messages in a transaction
      System.out.println("Session Transacted: " + session.getTransacted());
      sendMessage("First Message");
```

‡ This is called "transaction chaining," which means that the end of one transaction automatically starts another.

```
      sendMessage("Second Message");
      sendMessage("Third Message");
      session.commit();
      connection.close();
    } catch (Exception ex) {
      try {
        System.out.println("Exception caught, rolling back session");
        session.rollback();
      } catch (JMSException jmse) {
        jmse.printStackTrace();
      }
    }
  }

  private void sendMessage(String text) throws Exception {
    //send a simple text message within the group of messages
    TextMessage msg = session.createTextMessage(text);
    sender.send(msg);
  }  public static void main(String[] args) {
    try {
      JMSSenderTransacted app = new JMSSenderTransacted();
      app.sendMessages();
      System.exit(0);
    } catch (Exception up) {
        up.printStackTrace();
    }
  }

  public JMSSenderTransacted() {
    try {
      //create the connection, session, and sender
      Context ctx = new InitialContext();
      QueueConnectionFactory factory = (QueueConnectionFactory)
         ctx.lookup("QueueCF");
      connection = factory.createQueueConnection();
      connection.start();
      session =
         connection.createQueueSession(true, Session.AUTO_ACKNOWLEDGE);
      Queue queue = (Queue)ctx.lookup("queue1");
      sender = session.createSender(queue);
    } catch (Exception jmse) {
       jmse.printStackTrace();
       System.exit(0);
    }
  }
}
```

The `main` method instantiates a new `JMSSenderTransacted` object and invokes the `sendMessages` method, which sends a group of messages within the context of a `QueueSession` transaction. The constructor of this class establishes a connection to the JMS provider and sets up the JMS objects needed by the rest of the class.

The two key methods of this class related to JMS transactions are when the session is created and when the messages are sent:

```
public JMSSenderTransacted() {
  try {
    ...
    session =
       connection.createQueueSession(true, Session.AUTO_ACKNOWLEDGE);
  ...
}

public void sendMessages() {
  try {
    //send the messages in a transaction
    System.out.println("Session Transacted: " + session.getTransacted());
    sendMessage("First Message");
    sendMessage("Second Message");
    sendMessage("Third Message");
    session.commit();
    connection.close();
  } catch (Exception ex) {
    try {
      System.out.println("Exception caught, rolling back session");
      session.rollback();
     } catch (JMSException jmse) {
      jmse.printStackTrace();
    }
  }
}
```

Notice that when creating the `QueueSession`, a value of `true` is specified in the `createQueueSession` method. This specifies that the session should be transacted and all messages withheld from delivery to the message consumer until the session is committed. In the `sendMessages` method, three messages are sent, then the `commit` method on the `session` is invoked. At this point, all three messages are released by the JMS provider and are available for receipt by the message consumer.

If the `session.commit()` method is not invoked and the `sendMessages` method terminates normally, the messages will not be sent. No exception will be thrown and there is no record of the messages—they simply cease to exist without having been sent. It is therefore very important to ensure that the method logic always either commits the session or invokes the `rollback()` method.

If an exception occurs while sending the messages, say after `sendMessage("Second Message")`, then the `session.rollback()` method is invoked, rolling back all of the messages sent since the session was created.

Transacted sessions do not require the use of the JTA or an external container-based transaction manager to work properly. The transaction management is handled by the JMS provider. However, if you need to coordinate a single transaction across multiple sessions or between queues, topics, and a database, then transacted sessions are not enough. In this case, you will need to use JTA distributed transactions, which is discussed in the next section.

Distributed Transactions

Distributed systems sometimes use a *two-phase commit* (2PC) process that allows multiple distributed resources to participate in one transaction. This typically involves an underlying transaction manager that takes care of coordinating the prepare, commit, or rollback of each resource participating in the transaction. In most cases, the resources involved in the transaction are databases, but they can be other things, like JMS providers.

Transactions can be either local transactions or global transactions. Local transactions involve work performed on a single resource: one database or JMS provider. Global transactions involve work performed across several different resources (i.e., some combination of databases and JMS providers). JMS provides transaction facilities for both local and global transactions. The transacted sessions discussed previously are local transactions in JMS; they involve a single JMS provider.

The 2PC protocol is designed to facilitate global transactions—transactions that span multiple resources. As an example, an enterprise application may need to process (consume and produce) messages as well as make changes to a database. In some cases, the processing of messages and database updates needs to be treated as a single unit of work so that a failure to update the database or consume a message will cause the entire unit of work to fail. This is the basic premise behind a transaction: all the tasks must complete or fail together. To create a unit of work that spans different resources, the resources must be able to cooperate with a transaction manager in a 2PC.

The 2PC protocol is used by a transaction manager to coordinate the interactions of resources in a global transaction. A resource can only participate in a global transaction if it supports the 2PC protocol, which is usually implemented using the eXtended Architecture (XA) interface developed by The Open Group. In the Java enterprise technologies, the XA interface is implemented by the JTA and XA interfaces (`javax.transaction` and `javax.transaction.xa`). Any resource that implements these interfaces can be enrolled in a global transaction by a transaction manager that supports these interfaces.

JMS providers that implement the JTA XA APIs can therefore participate as a resource in 2PC. The JMS specification provides XA versions of the following JMS objects: `XAConnectionFactory`, `XAQueueConnection`, `XAQueueConnectionFactory`, `XAQueueSession`, `XASession`, `XATopicConnection`, `XATopicConnectionFactory`, and `XATopicSession`.

Each of these objects works like its corresponding non-XA-compliant object. The `XATopicSession`, for example, provides the same methods as the `TopicSession`. An application server's transaction manager uses these XA interfaces directly, but a JMS client only sees the nontransactional versions.

All XA-compliant resources (JDBC or JMS) provide an `XAResource` object that is an interface to the underlying resource (in JMS, the JMS provider). The `XAResource` object is used by the `TransactionManager` to coordinate the 2PC commit. In the previous example, the application associates the `XAResource` for the JDBC driver and the JMS provider with the current transaction so that all the work performed using those resources is bound together in one transaction. When the transaction is committed, all the work performed by the JDBC connection and JMS session is committed. If the transaction had been rolled back, all the work performed by the JDBC connection and JMS session would have been rolled back. All the work performed across these two resources either succeeds together or fails together.

An application server, such as an EJB server, may itself be a JMS client. In this case, whether the interfaces are exposed depends on how the JMS server and the application server are integrated. If the integration is hidden within the implementation, as is the case with EJB, then the container may use the XA-compliant version of these objects directly. Since the XA interfaces in JMS are not intended for application developers—they are intended to be implemented by vendors—we will not go into them in detail in this book. The important thing to understand is that JMS providers that implement the XA interfaces properly can be used in a 2PC transaction. If your application server (i.e., EJB server) supports 2PC, then these kinds of JMS providers can be used with other resources in global transactions.

Lost Connections

When the network connection between the client and server is lost, a JMS provider must make every reasonable attempt to reestablish the connection. In the event that the JMS provider cannot automatically reconnect, the provider must notify the client of this condition by throwing an exception when the client invokes a method that would cause network traffic to occur. However, it is reasonable to expect that a JMS client may only be a receiver using the `MessageListener` interface, and not a producer that makes any outbound `publish()` or `send()` calls. In this case, the client is not invoking JMS methods—it is just listening for messages—so a dropped connection may not be detected.

JMS provides an `ExceptionListener` interface for trapping a lost connection and notifying the client of this condition. The `ExceptionListener` is bound to the connection—unlike `MessageListeners`, which are bound to sessions. The definition of the `ExceptionListener` is:

```
public interface ExceptionListener {
   void onException(JMSException exception);
}
```

It is the responsibility of the JMS provider to call the `onException()` method of all registered `ExceptionListeners` after making reasonable attempts to reestablish the connection automatically. The JMS client can implement the `ExceptionListener` so that it can be alerted to a lost connection and possibly attempt to reestablish the connection manually.

How can the JMS provider call an ExceptionListener if the client has been disconnected? Every JMS provider has some amount of functionality that resides in the client application. We have been referring to this as the *client runtime* portion of the JMS provider. It is the responsibility of the client runtime to detect the loss of connection and call the ExceptionListener.

The ExceptionListener Example

To illustrate the use of the ExceptionListener interface, we start by changing the formal declaration of the JMSReceiver class used in the previous section to implement the javax.jms.ExceptionListener interface:

```
public class JMSReceiver implements
    javax.jms.MessageListener,
    javax.jms.ExceptionListener
{
...
```

Next, we remove the connection setup information from the constructor and isolate it in its own method, establishConnection():

```
public JMSReceiver() {
   establishConnection();
}
```

The establishConnection() method sets up the connection and other JMS objects needed by the class and registers the class as an ExceptionListener on the connection:

```
public void establishConnection() {
    try {
      Context ctx = new InitialContext();
      QueueConnectionFactory factory = (QueueConnectionFactory)
         ctx.lookup("QueueCF");
      QueueConnection connection = factory.createQueueConnection();
      connection.setExceptionListener(this);
    ...
}
```

Last, but not least, is the implementation of the onException() listener method. Its task is to call the establishConnection() method again to reestablish a connection with the JMS provider:

```
public void onException (javax.jms.JMSException jmse)
{
  // Tell the user that there is a problem
  System.err.println ("\n\nThere is a problem with the connection.");
  System.err.println ("   JMSException: " + jmse.getMessage());

  System.err.println ("Please wait while the application tries to "+
                      "reestablish the connection...");
  connect = null;
  establishConnection();
}
```

When a connection is dropped and reestablished, all of the sessions, queues, and senders need to be reestablished in order for the application to continue normal processing. This is why we isolated all the connection logic in the `establishConnection()` method, so that it can be used during startup and reused if the connection is lost.

JMS does not define any reason codes for a dropped connection. However, a JMS provider may provide a finer level of granularity by defining reason codes. Depending on the host operating system's network settings, it may take a while for the provider to notice that a connection has been dropped. Some providers implement a ping capability as a configurable setting to detect a network loss.

Dead Message Queues

JMS provides mechanisms for guaranteed delivery of messages between clients, utilizing the mechanisms we have discussed in this chapter. However, there are cases where guaranteed delivery, acknowledgments, and transactional semantics are just not enough. Many conditions may cause a message to be undeliverable. Messages may expire before they reach their intended destination or messages are viewed by the provider as undeliverable due to some other reason such as a deployment configuration problem. A message need not have an expiration associated with it, which means it would never expire. Forever is a long time. Realistically, it would be more prudent if the JMS provider could notify an application if a message cannot be delivered within a reasonable amount of time.

Although these issues are not specifically addressed by the JMS specification, some messaging vendors have the notion of a "Dead Letter Queue" or "Dead Message Queue" to deal with messages that are deemed undeliverable.

The extent of Dead Message Queue (DMQ) support varies from vendor to vendor. In the simplest case, it is the responsibility of the messaging system to put all undeliverable messages in the DMQ and it is the responsibility of the application to monitor its contents. In addition, a JMS provider may support administrative events that notify the application when something is placed in the DMQ. The notification may go to the sender or it may go to a centralized management tool. A specialized JMS client may be written to receive all DMQ notifications.

A DMQ can be treated just like a normal queue in most respects; it can be consumed or it can be browsed. There is one respect in which a DMQ behaves differently from other queues: The destination of a message, as obtained via `Message.getJMSDestination()`, would be the original destination the message was intended for, not the DMQ. The message may also contain additional properties, such as a vendor-defined reason code indicating why the message was placed in the DMQ.

It's important to know whether the JMS provider you are using supports Dead Message Queues. If it does, and you don't provide the application support to monitor it and peel things from the DMQ in a timely fashion, then the DMQ may fill up over time without your knowledge.

CHAPTER 8
Java EE and Message-Driven Beans

Java EE Overview

The examples in the book so far have primarily been Java SE applications using an external JMS provider (e.g., ActiveMQ, WebSphere MQ, and SonicMQ, to name a few). We will now look at how JMS can be used within the Java Platform, Enterprise Edition (Java EE). Java EE is a specification that unites several other Java enterprise technologies, including JMS, into one complete platform. Java EE is built on several main components: Web Services, Enterprise JavaBeans (EJB), and Java Management Extensions (JMX). Many other technologies, such as JMS, JDBC, JPA, JSP, JavaMail, JTA, and JNDI are also included as services in Java EE. The Java Message Service actually has two roles in Java EE: it is both a service and the basis for a special enterprise bean type called a message-driven bean, or simply MDB.

Java EE provides applications with several advantages, including object lifecycle management, container-managed resources, simplified deployment, load balancing, and the ability to seamlessly define and access remote objects. The Java EE container manages several types of resources, including data sources, JTA UserTransactions, and the EJB SessionContext. In addition to these resources, the Java EE container also manages JMS resources, including JMS destinations (queues and topics) and the JMS connection factory.

Per the Java EE specification, every application server compliant with the Java EE 4 specification and higher is also required to be a JMS provider as well, meaning that within the Java EE container environment you do not need an external JMS provider to use JMS messaging. However, as you will see in Chapter 11, many times it is desirable to have an external messaging provider, even within a Java EE container environment.

The Java EE specification ensures a certain amount of portability between vendors. A Java EE application that runs on Vendor A's platform should, with a little work, be able to run on Vendor B's Java EE platform. As long as proprietary extensions are not used, web, enterprise bean, and application client components developed to the Java EE specification will run on any Java EE platform.

Enterprise JavaBeans

Although JavaServer Pages (JSP) and Servlets can act as a JMS producer and synchronous consumer, the real power of JMS messaging comes from within Enterprise JavaBeans. The EJB3 specification (JSR-220) provides a simplified model for the development of Enterprise JavaBeans that significantly reduces the effort involved in developing and deploying Enterprise JavaBeans.

There are three main types of bean components in the EJB 3.0 specification: session beans, message-driven beans, and mapped entity objects. *Session beans* model processes and act as server-side extensions of their clients (they can manage a client's session state). *Message-driven beans (MDB)* are JMS clients that can consume (and send) messages concurrently in a robust and scalable EJB environment (these types of beans are the primary focus of this chapter). *Mapped entity objects* replace the older EJB 2.1 Entity Beans and are part of the Java Persistence API (JPA). They are used to model persistent business objects using well-known object-relational mapping (ORM) frameworks such as Hibernate, TopLink, and OpenJPA.

Application developers create custom enterprise beans by applying annotations to regular Java objects (or in the case of EJB 2.1, implementing one of the main bean interfaces) and developing the bean according to conventions and policies dictated by the EJB specification. Session beans model business processes that may or may not have session state. Session beans might be used to model business concepts such as a securities broker, an online shopping cart, loan calculation, medical claims processor—any process or mediator-type business concept. MDBs are used to model stateless JMS consumers. An MDB will have a pool of instances at runtime, each of which is a `MessageListener`. The bean instances can concurrently consume hundreds of messages delivered to the MDB, which makes the MDB scalable. Similar to session beans, message-driven beans model business processes by coordinating the interaction of other beans and resources according to the messages received and their payloads.

The EJB3 specification provides a rich set of Java metadata annotations that allow the bean developer to declare many of a bean's runtime behaviors including transaction policies, access control policies, and the resources (services) available. Resources (JMS, JDBC, JavaMail, etc.) that are declared through annotations are accessed via JNDI using a dependency injection annotation or alternatively from the bean's environment naming context (ENC). The ENC is a default read-only JNDI namespace that is available to every bean at runtime. Each bean deployment has its own JNDI ENC. In addition to providing a bean with access to resources such as JDBC, JavaMail, JTA, and URL and JMS connection factories, the JNDI ENC is used to access properties and other enterprise beans. Resources accessed from the JNDI ENC are managed implicitly by the EJB server so that they are pooled and then automatically enrolled in transactions as needed.

All enterprise beans (session, message-driven, and mapped entity) can be developed separately, packaged in a JAR file, and distributed. As components, packaged beans can be reused and combined with various other beans to solve any number of application requirements. In addition, enterprise beans are portable so that they can be combined and deployed on any application server that is EJB-compliant (providing you do not use any vendor-specific extensions).

Session beans are accessed as distributed objects via Java RMI-IIOP, which provides some level of location transparency; clients can access the beans on the server somewhat like local objects. Session beans are based on the RPC distributed computing paradigm. Message-driven beans are JMS clients that process JMS messages; they are *not* accessed as distributed objects. Message-driven beans are based on the asynchronous enterprise messaging paradigm.

Some of the details surrounding the EJB3 (JSR-220) specification are described in the next section. You can learn more about EJB by reading *Enterprise JavaBeans 3.0,* Fifth Edition (*http://oreilly.com/catalog/9780596009786/*), by Richard Monson-Haefel (O'Reilly).

Enterprise JavaBeans 3.0 (EJB3) Overview

The release of the EJB3 (JSR-220) specification in 2006 marked a significant turning point for EJB and Java EE in general. The primary theme of EJB3 is simplification and it achieves this goal very nicely. In addition to simplifying the development and deployment of Enterprise JavaBeans, the EJB3 specification added a new ORM-based persistence framework called JPA aimed at replacing the EJB 2.1 Entity Beans (which are no longer supported in EJB3).

One of the goals of the EJB3 specification is to address several issues found in the EJB 2.1 specification. Some of these issues included the dependency on a framework for the development of Enterprise JavaBeans (i.e., implementing and extending specific EJB interfaces), the need for home and remote interfaces, verbose and complex XML deployment descriptors, and the Entity Bean persistence model.

One of the significant productivity improvements in EJB3 is the use of Java metadata annotations over XML deployment descriptors (e.g., *ejb-jar.xml*). Of course, you can still choose to use XML deployment descriptors rather than annotations or you can use both XML and annotations together. When used in conjunction with annotations, the XML deployment descriptor overrides any matching configuration specified by the metadata annotation.

Other major features of the EJB3 framework include simplified bean development, the use of dependency injection, simplified callback methods for session and message-driven beans, the use of programmatic defaults that actually make sense, interceptors, and finally, the JPA. A description of each of these features is discussed in the following sections.

Simplified Bean Development

In EJB 2.1, an Enterprise JavaBean was made up of three components: a home interface, a remote interface, and the bean class. One of the issues with this model was that the home interface and remote interface were decoupled from the bean class and not associated with the bean class until the EJB stubs and skeletons were generated via the `ejbc` compile process. The relationship between the home interface, remote interface, and bean class was specified in the XML deployment descriptor (e.g., *ejb-jar.xml*):

```
<ejb-jar>
  <enterprise-beans>
    <session>
      <ejb-name>LenderBean</ejb-name>
      <home>LenderHome</home>
      <remote>Lender</remote>
      <ejb-class>LenderBean</ejb-class>
      <session-type>Stateless</session-type>
      <transaction-type>Container</transaction-type>
    </session>
  ...
  </enterprise-beans>
</ejb-jar>
```

The EJB3 specification addressed this issue by eliminating remote and home interfaces. Rather, the business interfaces are specified in the bean class through either the `@Remote` or `@Local` annotation. In addition, you are no longer required to implement the Session bean interface in your bean class. Session beans are developed as Plain Old Java Objects (POJOs) that either implement a corresponding business interface (or multiple business interfaces) specified through the `@Remote` or `@Local` annotation. Session beans are specified using the `@Stateless` or `@Stateful` annotations:

```
@Stateless
@Remote(Lender.class)
public class LenderBean implements Lender {
   ...
}
```

Notice that the `LenderBean` stateless session bean is defined only as a POJO with two annotations: one to define the object as a stateless session bean and the other to define which POJO interface should act as the remote interface for this bean. The `Lender.class` interface specified in the `@Remote` annotation is a plain Java interface. The nice thing about EJB3 is that no XML deployment descriptors are required to further define or deploy this bean. In addition, you no longer need to specify the `RemoteException` exception.

Dependency Injection

EJB3 offers limited dependency injection, allowing you to inject container managed resources and other Enterprise JavaBeans within an Enterprise JavaBean. It is limited because you can only inject *container-managed resources* into an Enterprise JavaBean.

Unlike other frameworks, EJB3 does not allow dependency injection into POJOs. One nice thing about EJB3 is that no corresponding setter method is required when defining an injected resource. To inject a container-managed resource, you would use the @Resource annotation followed by the JNDI name and the attribute declaration. Resources that can be injected into a bean include a data source, a JTA UserTransaction, a SessionContext, a JMS destination (e.g., Queue, Topic), and a JMS connection factory. To inject a bean into another bean for inter-service communication, you would use the @EJB annotation:

```
@Stateless
@Remote(Lender.class)
public class LenderBean implements Lender {

   @Resource SessionContext ctx;
   @Resource(name="jdbc/MasterDS") DataSource ds;
   @EJB protected CreditCheck creditCheck;
   ...
}
```

Once we get a reference to the LenderBean class, we have access within the bean to these resources. Notice that no corresponding setter method is required for these injected resources. We will be looking more at the @Resource annotation and how it relates to JMS later in this chapter.

Simplified Callback Methods

Because EJB3 session beans do not implement specific EJB framework components (e.g., javax.ejb.SessionBean), you are no longer required to override the annoying and rarely used callback methods (i.e., ejbActivate(), ejbPassivate(), ejbRemove(), and setSessionContext()). Instead, you can use corresponding annotations to annotate any method on the bean to be invoked for the specific callback desired. For stateless session beans and message-driven beans, the lifecycle callback annotations available are @PostConstruct and @PreDestroy. For stateful session beans, the lifecycle callback annotations available are @PostConstruct, @PreDestroy, @PostActivate, and @PrePassivate. You can apply these lifecycle callback annotations to any business method in the bean.

Programmatic Defaults

EJB3 makes use of *programmatic defaults* to simplify the development and deployment of Enterprise JavaBeans. For example, in EJB3 you no longer need to specify the EJB name for a bean; unless specified by the developer, the EJB name will automatically default to the fully qualified class name. In most cases, the fully qualified class name of the bean is unique enough to serve as the EJB name. In cases where it is not, you can always specify a name as an argument in the @Stateless annotation. The same holds true for the JNDI binding name—unless specified by the developer, each EJB3 container will automatically provide a default value for the JNDI binding name used to access the bean. For example, JBoss provides the following programmatic default for accessing a session bean from a servlet, JSP, or application client:

```
<earfile name>/<bean name>/<interface type>
```

The interface type is specified as either `remote` or `local`. For instance, to access a remote stateless session bean defined as `LenderBean` located in an EAR file named *example.ear*, you would specify the lookup as follows:

```
...
Context ctx = new InitialContext();
Lender Lender = (Lender)
  ctx.lookup("example/LenderBean/remote");
...
```

The client code just listed does not need a corresponding extended deployment descriptor (e.g., *jboss.xml*) to define the JNDI bindings. Using programmatic defaults, we know the EJB name (i.e., the fully qualified class name of the bean) and the JNDI binding (i.e., *example/LenderBean/remote*).

Interceptors

Interceptors are EJB3's answer to aspect-oriented programming (AOP). Unlike other aspect-oriented languages (e.g., AspectJ, Spring AOP), EJB3 interceptors do not require a specific aspect-oriented language. Rather, they use Java to define the interceptor method and annotations to specify the directives. While EJB3 interceptors are very powerful and easy to use, there are some limitations imposed on the use of interceptors. For example, you can use interceptors only on Enterprise JavaBeans; regular POJOs cannot be intercepted using EJB3 interceptors. Also, only one interceptor method can be defined per class, and a business method in a bean cannot be defined as an interceptor.

There are four annotations used for interceptors. The `@AroundInvoke` annotation defines a method as an interceptor. The rest of the annotations define the directives for the interceptor. The `@Interceptors` annotation, which is specified at a class level, defines the interceptors that are used for all methods in that bean class. The `@ExcludeClassInterceptors` annotation is used at the method level to indicate that the interceptors defined for the bean class should not be executed for a particular method. Finally, the `@ExcludeDefaultInterceptors` annotation is used at the class or method level to indicate that default interceptors should be excluded from the bean class and/or method. Default interceptors are interceptors defined in XML that apply to all bean classes.

To illustrate the use of interceptors in conjunction with JMS, consider the use case where any exception thrown by the backend session beans must publish that exception to an exception topic. That exception may then be picked up by anyone subscribing to that exception topic, including on-call systems, monitoring systems, and logging systems. Rather than adding the JMS publishing logic to every piece of code that throws an exception, we can define an interceptor to perform this cross-cutting concern:

```
public class ExceptionAOP {

   @AroundInvoke
   public Object sendException(InvocationContext ctx)
         throws Exception {
      String bean = ctx.getBean().getClass().getName();
      String op = bean + "->" + ctx.getMethod().getName();

      try {
         return ctx.proceed();
      } catch (Exception up) {
         InitialContext ictx = new InitialContext();
         TopicConnectionFactory factory =
            (TopicConnectionFactory)ictx.lookup("ConnectionFactory");
         TopicConnection conn = factory.createTopicConnection();
         Topic topic = (Topic)ictx.lookup("topic/exceptionTopic");
         TopicSession session =
            conn.createTopicSession(false, Session.AUTO_ACKNOWLEDGE);
         conn.start();

         TopicPublisher publisher = session.createPublisher(topic);
         TextMessage msg = session.createTextMessage(op + ":" + up);
         publisher.publish(msg);
         session.close();
         conn.close();
         throw up;
      }
   }
}
```

Notice that we can intercept the method and perform logic before the method is invoked as well as after the method is invoked. The `proceed()` method from the `InvocationContext` object is what triggers the invocation of the method being invoked. All interceptors must use the following method signature: `public Object myMethodName(InvocationContext ctx)`, where `myMethodName` is any method name you want (in our case it is `sendException`). The code to specify how this interceptor should be used is as follows:

```
@Stateless
@Remote(Lender.class)
@Interceptors(ExceptionAOP.class)
public class LenderBean implements Lender{

   ...
}
```

Notice that we specified via the `@Interceptors` annotation that the `ExceptionAOP` class (which is just a POJO) should be used as an interceptor for this bean class. You can chain interceptors by listing them as a class array in the `@Interceptors` annotation.

Java Persistence API

Finally, Entity beans in EJB 2.1 have been replaced with a new object-relational mapping (ORM) framework called the Java Persistence API (JPA). The great thing about JPA is that it does not require EJB; JPA can be used within Java EE or standalone through a Java SE application. The Spring Framework by SpringSource also provides seamless integration with JPA. JPA is a standard API and, as such, requires a JPA provider. Popular JPA providers include Hibernate, TopLink, and OpenJPA. You can read more about JPA in *Enterprise JavaBeans 3.0*, Fifth Edition (*http://oreilly.com/catalog/9780596009786/*), by Richard Monson-Haefel (O'Reilly).

JMS Resources in Java EE

The two JMS resources that must be obtained from the JNDI context are the JMS destination (`Queue` and `Topic`) and the JMS connection factory (`TopicConnectionFactory` and `QueueConnectionFactory`). In previous examples in this book, these resources were obtained by first getting an `InitialContext` to the JMS provider and then performing a lookup using the published JNDI name. Keeping with our borrower and lender example from the previous chapters, let's assume that we have a topic connection factory defined in the Java EE application server named "TopicCF," and a topic used to publish prices also defined in the Java EE application server named "jms/Rates." In previous examples, the constructor of the `Lender` class was where we established an `InitialContext` and obtained the `TopicConnectionFactory` and `Topic` via a JNDI lookup:

```
public class Lender {

   TopicConnection conn = null;
   TopicSession session = null;
   Topic ratesTopic = null;
   TopicPublisher publisher = null;

   public Lender() {
     try {

        Context ctx = new InitialContext();

        TopicConnectionFactory factory = (TopicConnectionFactory)
           ctx.lookup("TopicCF");
        conn = factory.createTopicConnection();
        conn.start();

        session =
           conn.createTopicSession(false,Session.AUTO_ACKNOWLEDGE);

        ratesTopic = (Topic)ctx.lookup("jms/Rates");
        publisher = session.createPublisher(ratesTopic);

        ...
```

```
        } catch (JMSException jmse) {
           jmse.printStackTrace();
        } catch (NamingException jne) {
           jne.printStackTrace();
        }
     }
     ...
}
```

Within the Java EE environment, objects bound within the Java EE context can use the `@Resource` annotation to inject the JMS resources (connection factories and destinations) into the code rather than performing a JNDI lookup using the `InitialContext`. Servlets, session beans, and message-driven beans can access JMS resources in this manner. Therefore, we can rewrite the previous Java SE code example in Java EE using resource injection as follows:

```
@Stateless
@Remote(Lender.class)
public class LenderBean implements Lender {

   @Resource(name="TopicCF") TopicConnectionFactory factory;
   @Resource(name="jms/Rates") Topic ratesTopic;

   TopicConnection conn = null;
   TopicSession session = null;
   TopicPublisher publisher = null;

   @PostContruct
   public void init() {
     try {
        conn = factory.createTopicConnection();
        conn.start();
        session =
           conn.createTopicSession(false,Session.AUTO_ACKNOWLEDGE);

        publisher = session.createPublisher(ratesTopic);
        ...

     } catch (JMSException jmse) {
        jmse.printStackTrace();
     }
   }
   ...
}
```

Notice that we do not need to establish an `InitialContext` and perform a JNDI lookup to obtain the JMS connection factory and topic resources accessed through JNDI; rather, they are injected through the `@Resource` annotation and are available to us immediately in the `init()` method.

The JNDI Environment Naming Context (ENC)

In older versions of Java EE (e.g., Java EE 4 and EJB 2.1), the JNDI Environment Naming Context (ENC) is used to access JNDI-based managed resources. The JNDI ENC specifies that JMS connection factories (TopicConnectionFactory and QueueConnectionFactory) can be bound within a JNDI namespace and made available to any Java EE component at runtime. This allows any Java EE component to become a JMS client.

For example, the Lender JMS client developed in Chapter 4 could be modeled as a Java EE application client, which would allow it to access a JMS connection factory through the JNDI ENC:

```
public class Lender implements javax.jms.MessageListener {

   TopicConnection conn = null;
   TopicSession session = null;
   TopicPublisher publisher = null;

   public Lender() {
      try{
         InitialContext jndiEnc = new InitialContext();

         TopicConnectionFactory factory = (TopicConnectionFactory)
            jndiEnc.lookup("java:comp/env/jms/TopicCF");
         connect = factory.createTopicConnection();

         session =
          connect.createTopicSession(false,Session.AUTO_ACKNOWLEDGE);

         Topic ratesTopic = (Topic)
            jndiEnc.lookup("java:comp/env/jms/Rates");
         publisher = session.createPublisher(ratesTopic);

         ...

      } catch (javax.jms.JMSException jmse) {
         jmse.printStackTrace(); System.exit(1);
      } catch (javax.naming.NamingException jne) {
         jne.printStackTrace(); System.exit(1);
      }
   }
...
```

Notice that the InitialContext did not need a set of vendor-specific properties and that the lookup() operations referenced a special namespace, "java:comp/env/jms/", to access the connection factories. The JNDI ENC allows the Java EE component to remain ignorant of the actual network location of the administered objects, and even of the vendor that implements them. This allows the Java EE components to be portable across JMS providers as well as Java EE platforms. In addition, the JNDI names used to locate objects are logical bindings, so the topics or queues bound to these names can change independently of the actual bindings used by the JMS provider.

In the XML deployment descriptor for the Lender application client, the component developer declares that a JMS connection factory and destination need to be bound within the JNDI ENC:

```
<application-client>
  <display-name>Lender Applicaton</display-name>
    <resource-ref>
      <description>Lender Broker</description>
      <res-ref-name>jms/TopicCF</res-ref-name>
      <res-type>javax.jms.TopicConnectionFactory</res-type>
      <res-auth>Container</res-auth>
    </resource-ref>
    ...
    <resource-env-ref>
      <description>Rates Topic</description>
      <resource-env-ref-name>jms/Rates</resource-env-ref-name>
      <resource-env-ref-type>javax.jms.Topic</resource-env-ref-type>
    </resource-env-ref>
    ...
</application-client>
```

When the component is deployed, the Java EE vendor tools generate code to translate the JNDI ENC resource references into JMS-administered objects. This translation is done when the bean is deployed using administration tools.

Any J2EE component can access JMS connection factories and destinations using the JNDI ENC. As an example, the Lender client can be rewritten as a stateless session bean that uses the JNDI ENC to obtain a JMS connection factory and destination:

```
public class LenderBean implements javax.ejb.SessionBean {

  ...

  public void setSessionContext(SessionContext cntx) {
    try {

      InitialContext jndiEnc = new InitialContext();

      TopicConnectionFactory factory = (TopicConnectionFactory)
        jndiEnc.lookup("java:comp/env/jms/TopicCF");
      connect = factory.createTopicConnection();

      session =
       connect.createTopicSession(false,Session.AUTO_ACKNOWLEDGE);

      Topic RatesTopic=(Topic)
          jndiEnc.lookup("java:comp/env/jms/Rates");
      publisher = session.createPublisher(ratesTopic);

      ...

    }
    ...
```

```
    }
    public void publishRates(double oldRate, double newRate) {
       try {
         javax.jms.StreamMessage message =
                   session.createStreamMessage();
         message.writeDouble(oldRate);
         message.writeDouble(newRate);
         ...

         publisher.publish(
             message,
             DeliveryMode.PERSISTENT,
             Message.DEFAULT_PRIORITY,
             1800000);

       } catch ( javax.jms.JMSException jmse ){
          jmse.printStackTrace();
       }
    }
    ...
}
```

Message-Driven Beans

Although session beans and web components can act as JMS producers, these components can only consume JMS messages *synchronously* using one of the `MessageConsumer.receive()` methods. Calling the `receive()` methods causes the JMS client to block on the currently running thread and wait for a message.* These methods are used to consume messages synchronously, whereas `MessageListener` objects are used to consume messages asynchronously.

Only the message-driven bean and application client components can both produce and consume asynchronous messages. The web and session components cannot act as asynchronous JMS consumers because they are driven by synchronous request/reply protocols, not asynchronous messages. Web components respond to HTTP requests while session beans respond to Java RMI-IIOP requests.

The fact that neither web components nor session beans can asynchronously consume JMS messages was one of the things that originally led to development of the message-driven bean. The MDB provides Java EE developers with a server-side JMS consumer that can consume asynchronous messages, something that wasn't supported in the early versions of Java EE.

* It's recommended that the component developer use the timeout version of the receive method, `receive(long timeout)`, where the timeout is specified in milliseconds. Unrestricted blocking not limited to any length of time is risky.

While most JMS vendors provide the message-brokering facilities for routing messages from producers to consumers, the responsibility for implementing JMS clients is left to the application developer. In many cases, the JMS clients that consume and process messages need a lot of infrastructure in order to be robust, secure, fault-tolerant, and scalable. JMS clients may access databases and other resources, use local and distributed transactions, require authentication and authorization security, or need to process a large load of concurrent messages. Fulfilling these needs is a tall order, requiring that a significant amount of work be done by the application developer. In the end, the kind of infrastructure needed to support powerful JMS consumers is not unlike the infrastructure needed for session beans, which can produce but not consume messages asynchronously.[†]

In recognition of this need, Java EE provides support for the `MessageDrivenBean` type, which can consume JMS messages and process them in the same robust component-based infrastructure that session beans enjoy. The `MessageDrivenBean` type (message-driven bean) is a first-class enterprise bean component that is designed to consume asynchronous JMS messages. Like stateless session beans, message-driven beans don't maintain state between requests; they may also have instance variables that are maintained throughout the bean instance's life, but they may not store conversational state. Unlike session beans, a message-driven bean does not have remote or local business interfaces associated with them, because the MDB is not an RPC component. Furthermore, it does not have business methods that are invoked by EJB clients. An MDB consumes messages delivered by other JMS clients through a message server.

The lifecycle of a message-driven bean is shown in Figure 8-1. The `setMessageDrivenContext()` is called by the container on each instance right after it is instantiated. It provides the instance with a `MessageDrivenContext`, which is based on a standard container interface, `EJBContext`. The `ejbCreate()` method is invoked by the container on each instance after the `setMessageDrivenContext()` method, but before the bean instance is added to the pool for a particular message-driven bean. Once the MDB has been added to the pool, it's ready to process messages. When a message arrives, the instance is removed from the pool and its `onMessage()` method is invoked. When the `onMessage()` method returns, the bean instance is returned to the pool and is ready to process another message. The `ejbRemove()` method is invoked by the container on an instance when it is discarded. This might happen if the container needs to reduce the size of the pool.

[†] Session beans can technically consume JMS messages synchronously by using one of the `MessageConsumer.receive()` methods.

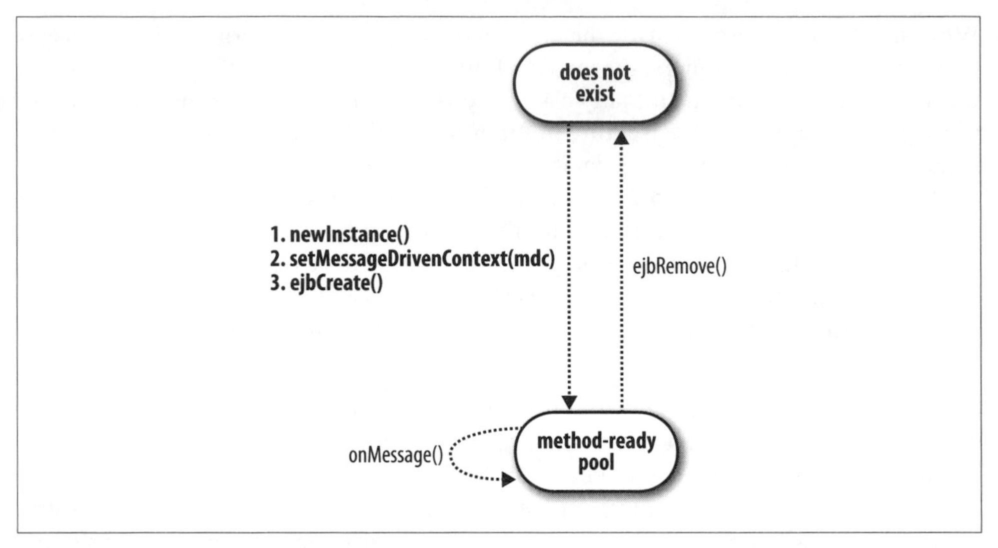

Figure 8-1. Lifecycle of a message-driven bean

Concurrent Processing and Scalability

In addition to providing the container infrastructure for message-driven beans, Java EE provides another important advantage: concurrent processing. A message-driven bean is deployed as a JMS consumer. It subscribes to a topic or connects to a queue and waits to receive messages. At runtime, the EJB container actually instantiates many instances of the same MDB and keeps those instances in pool. When a message is delivered to an MDB, one instance of that bean is selected from a pool to handle the message. If several messages are delivered at the same time, the container can select a different bean instance to process each message; the messages can be processed concurrently. Because a message-driven bean can consume messages concurrently in a robust server environment, it is capable of much higher throughput and better scalability than most traditional JMS clients.

Defining Message-Driven Beans

In EJB3, message-driven beans are defined through the `@MessageDriven` annotation. In addition, the bean class must implement the `javax.jms.MessageListener` interface. The `@MessageDriven` annotation is used to identify the bean as a message-driven bean, whereas the associated `@ActivationConfigProperty` annotation is used to specify the JMS destination the message-driven bean is listening on:

```
@MessageDriven(activationConfig = {
   @ActivationConfigProperty(
      propertyName="destinationType",
      propertyValue="javax.jms.Queue"),
   @ActivationConfigProperty(
```

```
        propertyName="destination",
        propertyValue="jms/LoanRequest")
})
```

Notice here that two properties are set; the first identifies the destination type (`Queue` or `Topic`), and the second identifies the JNDI name of the JMS destination. Like session beans, no XML deployment descriptors are required to develop and deploy a MDB. You can specify other MDB-related properties here as well, including a message selector and the number of concurrent listener threads.

The `MessageListener` interface defines the `onMessage()` method, which is used to asynchronously receive messages:

```
package javax.jms;

public interface MessageListener {
   public void onMessage();
}
```

Per the JMS specification, all MDBs must implement the `MessageListener` interface.

The `@PostConstruct` callback method (if specified) is called on each instance right after it is instantiated. Prior to the callback, the instance is provided with a `MessageDriven Context`, which is based on a standard container interface, `EJBContext`. Once the message-driven bean has been added to the pool, it's ready to process messages. When a message arrives, the instance is removed from the pool and its `onMessage()` method is invoked. When the `onMessage()` method returns, the bean instance is returned to the pool and is ready to process another message. The `@PreDestroy` callback method (if specified) is invoked on an instance when it is discarded. This might happen if the container needs to reduce the size of the pool.

The `Lender` JMS client developed in Chapter 4 can easily be converted to a message-driven bean. When messages are received from the borrowers, the `LenderMDB` class can process them quickly and efficiently, providing a more scalable option then the JMS clients we previously developed:

```
@MessageDriven(activationConfig = {
   @ActivationConfigProperty(
      propertyName="destinationType",
      propertyValue="javax.jms.Queue"),
   @ActivationConfigProperty(
      propertyName="destination",
      propertyValue="jms/LoanRequest")
})
public class LenderMDB implements MessageListener {

    @Resource(name="jms/QueueCF") private QueueConnectionFactory factory;

    private javax.jms.QueueConnection qConnect = null;
    private javax.jms.QueueSession qSession = null;

    @PostConstruct
    public void init(){
```

```
        try {
            //needed for request/reply processing
            qConnect = factory.createQueueConnection();
            qSsession = qConnect.createQueueSession
                      (false,Session.AUTO_ACKNOWLEDGE);
            qConnect.start();

        } catch (JMSException jmse){
            throw new RuntimeException();
        }
    }

    @PreDestroy
    public void cleanup(){
        try {
            qConnect.close();
        } catch (JMSException jmse){
            throw new RuntimeException();
        }
    }

    public void onMessage(Message message){
        try{
           boolean accepted = false;

           // Get the data from the message
           MapMessage msg = (MapMessage)message;
           double salary = msg.getDouble("Salary");
           double loanAmt = msg.getDouble("LoanAmount");

           // Determine whether to accept or decline the loan
           if (loanAmt < 200000) {
              accepted = (salary / loanAmt) > .25;
           } else {
              accepted = (salary / loanAmt) > .33;
           }
           System.out.println("" +
              "Percent = " + (salary / loanAmt) + ", loan is "
              + (accepted ? "Accepted!" : "Declined"));

           // Send the results back to the borrower
           TextMessage tmsg = qSession.createTextMessage();
           tmsg.setText(accepted ? "Accepted!" : "Declined");
           tmsg.setJMSCorrelationID(message.getJMSMessageID());

           // Create the sender and send the message
           QueueSender qSender =
              qSession.createSender((Queue)message.getJMSReplyTo());
           qSender.send(tmsg);

           System.out.println("\nWaiting for loan requests...");
        } catch (Exception rte) {
            throw new RuntimeException();
        }
```

```
        }
    }
```

The `LoanMDB` message-driven bean listens for loan requests on the `loanRequest` queue to determine whether the loan request should be approved or denied. The `QueueConnectionFactory` is obtained from the JNDI ENC using the `@Resource` annotation, eliminating the code necessary to obtain an initial context and get the connection factory. However, where is the code used to specify the `loanRequest` queue?

A message-driven bean is, by definition, a JMS consumer. The EJB container in which the bean is deployed takes care of registering the bean as a listener to the desired queue based on JMS configuration information provided by the developer or deployer. In the case of message-driven beans, the bean developer or deployer can use either the `@MessageDriven` annotation and the corresponding `@ActivationConfigProperty` annotation or the corresponding XML deployment descriptors. The `@MessageDriven` annotation includes properties for describing the type of destination (topic or queue), whether to use durable or nondurable subscriptions with topics, the acknowledgment mode, and even the message selector used.

With the message-driven bean, it is important to understand that messages do not have to be produced by other beans in order for them to be consumed by an MDB. Message-driven beans can consume messages from any topic or queue administered by a JMS provider.‡ Messages consumed by MDBs may have come from other beans (session beans or message-driven beans), web components, application client components, non-J2EE JMS clients, or even legacy messaging systems supported by a JMS provider. A legacy application might, for example, use IBM's WebSphere MQ to deliver messages to a queue, which is consumed by other legacy applications as well as message-driven beans.

Message-Driven Bean Use Cases

We saw earlier in the chapter that within the Java EE environment, session beans and web components can only receive messages synchronously, whereas message-driven beans can receive messages asynchronously. While this is a key point for the usefulness of MDBs, there are several other use cases that make MDBs a good design choice.

Message Facade

There may be times when you want to expose business functionality written in Java to remote client applications outside of the EJB container. This is typically done though the use of the Session Facade pattern. In Java EE, this pattern is implemented via stateless session beans that act as the remote entry point to the business functionality we

‡ In almost all cases, the EJB vendor will also be the JMS provider, provided the EJB container is Java EE 1.4 (and above) compliant.

want to expose. However, this implementation assumes that the client invoking the service is written in Java, can support the RMI/IIOP protocol, and has the necessary interfaces, stubs, and skeletons to access the remote session bean. While this solution is valid for Java-based client applications, it does not lend itself well to interoperability with non-Java clients or Java clients outside of the Java EE environment.

Another approach for exposing business functionality within an EJB container to outside client applications is through the use of the Message Facade pattern. Using a message-driven bean as a facade to business functionality (services) decouples client applications from the service implementation, allowing for greater flexibility and interoperability, particularly for non-Java client applications wishing to use the service. In this scenario, the message-driven bean acts as the facade to receive the request via a JMS message, invoke the necessary service, and return the result back to the caller in the form of a JMS message. Consider the example in Chapter 4 where the `Borrower` class makes a request to the `Lender` class for a mortgage loan. Rather than convert the entire `Lender` object to a message-driven bean, we could create a message facade (in this case `LenderMDB`) that handles the loan request:

```
@MessageDriven(activationConfig = {
   @ActivationConfigProperty(
      propertyName="destinationType",
      propertyValue="javax.jms.Queue"),
   @ActivationConfigProperty(
      propertyName="destination",
      propertyValue="jms/LoanRequest")
})
public class LenderMDB implements MessageListener {

    @Resource(name="jms/QueueCF") private QueueConnectionFactory factory;

    private javax.jms.QueueConnection qConnect = null;
    private javax.jms.QueueSession qSession = null;

    @PostConstruct
    public void init() {
        try {
            qConnect = factory.createQueueConnection();
            qSsession = qConnect.createQueueSession
                      (false,Session.AUTO_ACKNOWLEDGE);
            qConnect.start();

        } catch (JMSException jmse) {
            throw new RuntimeException();
        }
    }

    @PreDestroy
    public void cleanup(){
        try {
            qConnect.close();
        } catch (JMSException jmse) {
            throw new RuntimeException();
```

```
        }
    }

    public void onMessage(Message message) {
        try{
           // Get the data from the message
           MapMessage msg = (MapMessage)message;
           double salary = msg.getDouble("Salary");
           double loanAmt = msg.getDouble("LoanAmount");

           // Determine whether to accept or decline the loan
           Lender lender = new Lender();
           boolean accepted = lender.analyzeLoadRequest(salary, loanAmt);

           // Send the results back to the borrower
           TextMessage tmsg = qSession.createTextMessage();
           tmsg.setText(accepted ? "Accepted!" : "Declined");
           tmsg.setJMSCorrelationID(message.getJMSMessageID());

           // Create the sender and send the message
           QueueSender qSender =
              qSession.createSender((Queue)message.getJMSReplyTo());
           qSender.send(tmsg);

           System.out.println("\nWaiting for loan requests...");
        } catch (Exception rte) {
            throw new RuntimeException();
        }
    }
}
```

Notice that by using a message-driven bean as a message facade to the loan request, the `Lender` business object did not need to be converted to a remote object (i.e., session bean) or a JMS consumer and producer (i.e., message-driven bean). In this example, loan requests are received in the `jms/LoanRequest` queue in the form of a `MapMessage` containing the salary and loan amount information and returned via the `jms/LoanResponse` queue (specified in the `JMSReplyTo` message header property) as a `TextMessage` containing the information on whether the loan was approved or denied. The key point here is that the `LenderMDB` facade abstracts out the JMS infrastructure and connection logic from the core business logic, allowing the class containing the business logic to be independent of how the functionality is exposed (in our case, JMS).

Transformation and Routing

In the previous example, we saw how a message-driven bean can be used as a message facade to a POJO-based business object, allowing the POJO business functionality to be exposed outside of the EJB container. Suppose the client application making the loan request was written in a language other than Java? Suppose also that the `analyzeLoanRequest()` method in the `Lender` class took a `LoanRequest` object rather than two double arguments? One obvious solution to this problem would be to have the client application use XML to make the request. However, we would then have to modify the `analyzeLoanRequest()` method of the `Lender` class to accept XML rather than a `LoanRequest` object.

In this scenario, we can use a message-driven bean to act as a message facade but also have the additional responsibility of transforming the message to the appropriate contract definition language (in this case, going from XML to a Java object and back to XML for the response). Although the topic of message transformation is beyond the scope of this book, conceptually a message-driven bean can perform this task as follows:

```
@MessageDriven(activationConfig = {
   @ActivationConfigProperty(
      propertyName="destinationType",
      propertyValue="javax.jms.Queue"),
   @ActivationConfigProperty(
      propertyName="destination",
      propertyValue="jms/LoanRequest")
})
public class LenaderMDB implements MessageListener {

    ...

    public void onMessage(javax.jms.Message message) {
        try{
           // Get the data from the message
           BytesMessage msg = (BytesMessage)message;
           String xmlIn = msg.readUTF();

           LoanRequest loan = transformFromXML(xmlIn);

           Lender lender = new Lender();
           boolean accepted = lender.analyzeLoanRequest(loan);

           String xmlOut = transformToXML(accepted);

           BytesMessage returnMsg = session.createBytesMessage();
           returnMsg.writeUTF(xmlOut);
           sender.send(returnMsg);

        } catch (Exception e) {
            throw new RuntimeException();
        }
    }
}
```

Additionally, we could add further functionality to our message-driven bean to have it interrogate the message properties to determine the name of the operation to be invoked and effectively "route" the request (via method invocation) to the requested service. In effect, the message-driven bean performing this functionality is essentially doing what an Enterprise Service Bus (ESB) does: accept a request, transform the message, and route the request. Depending on the complexity of the application, with a message-driven bean you could conceivably build a simple mediator component that handles transformations, routing, and request management.

Although the actual development of message-driven beans with transformation and routing capabilities would be a complex undertaking, it nevertheless illustrates the possible use cases of message-driven beans within an overall integration architecture solution.

CHAPTER 9

Spring and JMS

The Spring Framework provides built-in support for JMS that greatly simplifies the development of messaging-based applications. For synchronous message sending and receiving, Spring provides a JMS template that abstracts the developer from the details of the JMS API. For asynchronous message receiving, Spring provides a framework that allows regular POJOs to act as asynchronous message listeners (also known as message-driven POJOs, or MDPs). Message-driven POJOs are similar in nature to the message-driven beans discussed in the previous chapter. However, unlike message-driven beans, MDPs can be created using standard POJO business objects that optionally have no knowledge of messaging or JMS.

This chapter introduces JMS messaging using version 2.5 of the Spring Framework. We will start by reviewing the overall architecture and components of Spring's messaging framework, and then describe the details of connecting to a JMS provider, sending messages, receiving messages synchronously, and creating message-driven POJOs to receive messages asynchronously. In the last section ("The Spring JMS Namespace" on page 208), we will discuss the new Spring JMS namespace, which simplifies the configuration for Spring-based JMS containers and listeners.

Spring Messaging Architecture

The *JMS template* and the *message listener container* are the two main components within the Spring Framework for using JMS messaging. Spring's JMS template (`JmsTemplate`) is used when sending messages or receiving messages synchronously (i.e., blocking when receiving messages). The message listener container (`DefaultMessageListenerContainer`) is used to receive messages asynchronously through the use of MDP. Unlike most Java EE application servers (e.g., JBoss, WebSphere), Spring itself is not a JMS provider, meaning that an external JMS provider (e.g., ActiveMQ, JBoss Messaging, IBM WebSphere MQ) is necessary to use messaging with Spring. The purpose of the JMS template and the message listener container is to isolate the developer from the details of connecting to the JMS provider, establishing a JMS connection, creating a JMS session, and creating a JMS message producer or message consumer. The high-level architecture of how Spring is used for messaging is illustrated in Figure 9-1.

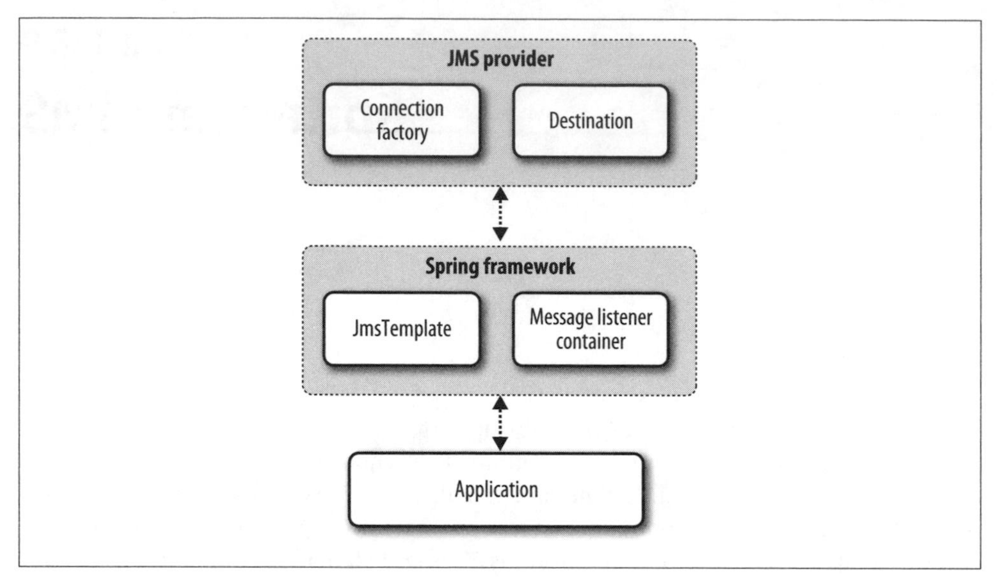

Figure 9-1. High-level architecture

The Spring JMS template (`JmsTemplate`) is the primary interface used for sending messages and receiving messages synchronously. When using JNDI, the `JmsTemplate` is used in conjunction with several other Spring objects to connect to the JMS provider, including the `JndiTemplate`, `JndiObjectFactoryBean`, `JndiDestinationResolver`, and the `CachingConnectionFactory` (or `SingleConnectionFactory`). The `JndiTemplate` bean is used to specify the initial factory, provider URL, security principals, and security credentials properties for connecting to the JMS provider. It is used when defining the JMS connection factory via the `JndiObjectFactoryBean` and the JMS destinations via the `JndiDestinationResolver`. These Spring beans are discussed in more detail in the section on connecting to a JMS provider. Figure 9-2 illustrates the relationship and collaboration among Spring's JNDI objects, the JMS provider, and the application.

For receiving messages asynchronously (i.e., listening for messages on a queue or topic without blocking), Spring provides a message listener container (`DefaultMessageListenerContainer` or `SimpleMessageListenerContainer`) that is used to create what is called a MDP. At first glance, MDPs seem similar to the message-driven beans found in the Java EE specification. However, Spring MDPs provide much more flexibility than MDBs in that Spring MDPs can be created from non-message-aware POJOs, whereas MDBs must adhere to specific rules in the EJB3 specification—namely, the object must implement the `javax.jms.MessageListener` interface, override the `onMessage` method, and provide a `@MessageDriven` annotation (or XML) containing configuration properties containing the destination type (e.g., `javax.jms.Queue` or `javax.jms.Topic`) and destination JNDI name.

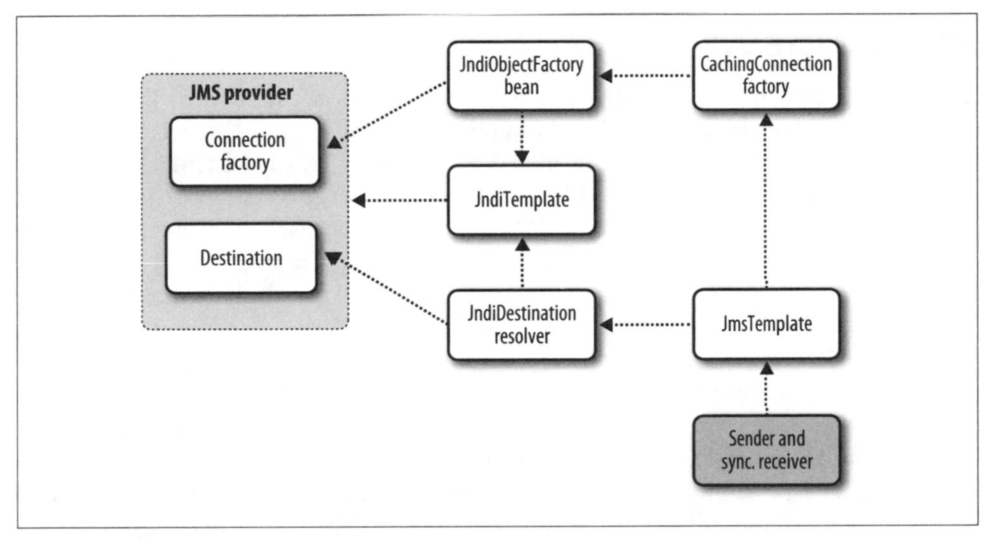

Figure 9-2. Synchronous send and receive using JNDI

Like the `JmsTemplate`, the message listener container is used in conjunction with the `JndiTemplate`, `JndiObjectFactoryBean`, `JndiDestinationResolver`, and the `CachingConnectionFactory` to connect to the JMS provider and start the asynchronous listener. Unlike message-driven beans, Spring provides three different ways to create MDPs. The three methods for creating a Spring MDP are outlined and discussed later in the section "Message-Driven POJOs" on page 198. Figure 9-3 illustrates the relationships and objects used for receiving asynchronous messages.

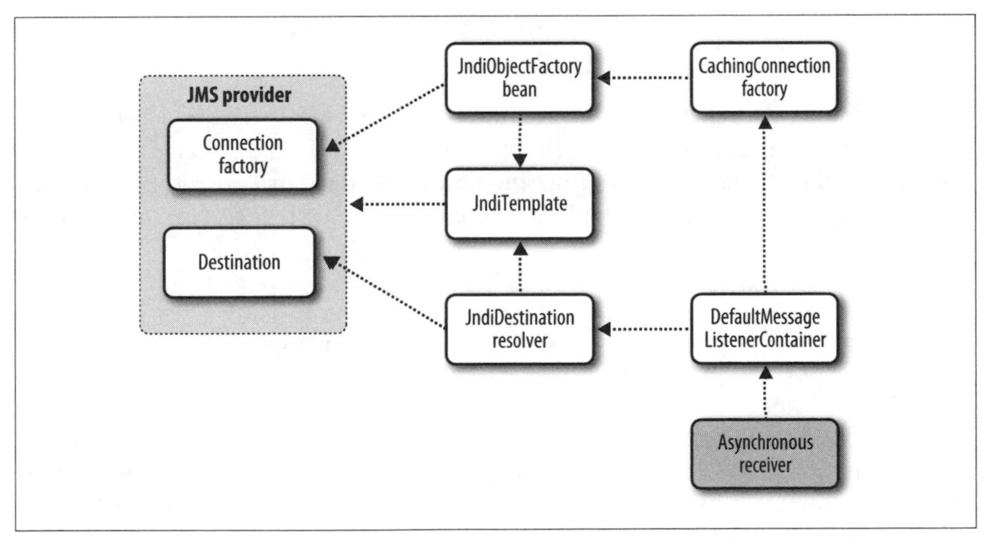

Figure 9-3. Asynchronous receive using JNDI

JmsTemplate Overview

The JmsTemplate is the primary object used for sending messages and receiving messages synchronously (i.e., blocking while waiting to receive messages). There is a JMS template version for JMS 1.1 (JmsTemplate) and a version for JMS 1.0.2 (JmsTemplate102). Since most JMS providers and Java EE application servers now support JMS 1.1, we will be focusing our attention on the JMS 1.1 Spring objects. Using the JmsTemplate significantly reduces the development effort involved in sending and receiving messages. When using the JmsTemplate, you do not need to worry about connecting to the JMS provider, creating a JMS session (e.g., QueueSession), creating a message producer (e.g., QueueSender), or even creating a JMS message (e.g., TextMessage). The JmsTemplate can automatically convert String objects, Byte[] objects, Java objects, and java.util.Map objects into the corresponding JMS message object types. You can also provide your own message converter to provide custom converters for complex messages or other types of messages not supported by the default message converter. The following code example illustrates the simplicity of sending a simple text message using Spring:

```
public class SimpleJMSSender {

   public static void main(String[] args) {
     try {
       ApplicationContext ctx =
           new ClassPathXmlApplicationContext("app-context.xml");
       JmsTemplate jmsTemplate =
           (JmsTemplate)ctx.getBean("jmsTemplate");

       jmsTemplate.convertAndSend("This is easy!");
     }
     ...
   }
}
```

In this example, the convertAndSend method will automatically create a TextMessage from the String argument containing the text "This is easy!" and send it to the default JMS destination (queue1) specified in the defaultDestinationName property of the jmsTemplate bean (we will be covering the configuration of the JmsTemplate in more detail in subsequent sections):

```
<bean id="jmsTemplate" class="org.springframework.jms.core.JmsTemplate">
   <property name="connectionFactory" ref="queueConnectionFactory"/>
   <property name="destinationResolver" ref="destinationResolver"/>
   <property name="defaultDestinationName" value="queue1"/>
</bean>
```

But wait a minute—what about connecting to the JMS provider, establishing a Queue Session, and creating a QueueSender? The JmsTemplate takes care of all these details (including cleanup), allowing you to focus on business logic rather than the JMS infrastructure. Of course, things can get a little more complex, particularly when you need to set message headers, message properties, or even create your own JMS message objects. We'll cover those details in subsequent sections of this chapter.

Out of the 75 methods defined in the `JmsTemplate` class (located in the `org.springframework.jms.core` package), the most common methods that you will likely use are `send`, `convertAndSend`, `receive`, `receiveSelected`, `receiveAndConvert`, and `receiveSelectedAndConvert`. Each of these common methods are outlined and discussed in the following subsections.

Send Methods

When using the `send` method in the `JmsTemplate` class, you will need to create an instance of Spring's `MessageCreator` class. This class, which is usually specified as an anonymous inner class within the `send` method parameter list, is used to create and populate a JMS `Message` object. The `send` method then invokes the `createMessage` method on the `MessageCreator` object to get the message to send.

When you have a default destination specified in the `JmsTemplate` and you wish to send a message to that default destination, you would use the following `send` method:

```
public void send(MessageCreator messageCreator)
                 throws JmsException
```

If you are not using a default destination in the `JmsTemplate` or you want to send to a destination other than the default destination specified, you can use the following overridden methods and specify a destination directly in the `send` method as a `String` argument containing the JNDI name or a JMS `Destination` argument (e.g., `javax.jms.Queue` or `javax.jms.Topic`):

```
public void send(String destinationName,
                 MessageCreator messageCreator)
                 throws JmsException

public void send(Destination destination,
                 MessageCreator messageCreator)
                 throws JmsException
```

We will cover the details of actually using these methods later in this chapter.

convertAndSend Methods

The `convertAndSend` methods will invoke a message converter to automatically convert a `String` object into a JMS `TextMessage`, a Java object into an JMS `ObjectMessage`, a `byte[]` array into a JMS `BytesMessage`, and a `java.util.Map` object into a JMS `MapMessage`. There are six variations of the `convertAndSend` method. The `convertAndSend` methods use a *message converter* to convert the `Object` argument into a JMS `Message`. If you do not specify a message converter, the `convertAndSend` methods will automatically use the default message converter provided by Spring (`SimpleMessageConverter`).

To have the `JmsTemplate` convert an object into a corresponding JMS message type using the default destination, you would use the following `convertAndSend` method:

```
public void convertAndSend(Object message)
                           throws JmsException
```

If you are not using a default destination in the `JmsTemplate` or you want to send to a destination other than the default destination specified, you can use the following overridden methods and specify a destination directly in the `convertAndSend` method as a `String` argument containing the JNDI name or a JMS `Destination` argument:

```
public void convertAndSend(String destinationName,
                           Object message)
                           throws JmsException

public void convertAndSend(Destination destination,
                           Object message)
                           throws JmsException
```

Finally, there may be times when you need to set some message headers or application properties on the message or perform other logic prior to the message being sent. To do this, you could use one of the overridden methods that take an instance of a `MessagePostProcessor` class as an argument, which is usually defined as an anonymous inner class. The use of the `MessagePostProcessor` will be covered in more detail in the following sections of this chapter. The following three overridden methods use the `MessagePostProcessor`:

```
public void convertAndSend(Object message,
                           MessagePostProcessor postProcessor)
                           throws JmsException

public void convertAndSend(String destinationName,
                           Object message,
                           MessagePostProcessor postProcessor)
                           throws JmsException

public void convertAndSend(Destination destination,
                           Object message,
                           MessagePostProcessor postProcessor)
                           throws JmsException
```

receive and receiveSelected Methods

The `receive` methods are used to block and wait for a message from a specified queue or topic. You can set the receive timeout by setting the `receiveTimeout` property on the `JmsTemplate` bean or by invoking the `setReceiveTimeout` method directly. The six overridden `receive` methods all return a `javax.jms.Message` object, which can then be cast into one of the five JMS message types. If you want to receive all messages from a default destination, you can use the following `receive` method:

```
public Message receive() throws JmsException
```

If you do not have a default destination defined or wish to receive all messages from another destination, you can use either of the following receive methods, passing in a String argument containing the JNDI name of the destination or a reference to a JMS destination:

```
public Message receive(String destinationName)
                       throws JmsException

public Message receive(Destination destination)
                       throws JmsException
```

Each of the receive methods just shown has a corresponding receiveSelected method that accepts a message selector as a String object. You can use these if you want to be more selective about the messages you want to receive:

```
public Message receiveSelected(String messageSelector)
                               throws JmsException

public Message receiveSelected(String destinationName,
                               String messageSelector)
                               throws JmsException

public Message receiveSelected(Destination destination,
                               String messageSelector)
                               throws JmsException
```

receiveAndConvert Methods

As with the convertAndSend methods, Spring provides the ability to convert the JMS Message object to a corresponding Java object type upon receipt of the message. Each of the receiveAndConvert methods returns a Java object based on the message type. Using the default message converter, the receiveAndConvert method will return a String object when receiving a TextMessage, a byte[] array when receiving a BytesMessage, a java.util.Map object when receiving a MapMessage, and finally a Java Object when receiving an ObjectMessage. There are three variations of the receiveAndConvert method: one that uses the default destination, one that accepts a String argument containing the JNDI name of the JMS destination, and one that accepts a Destination object containing the queue or topic from which to receive messages:

```
public Object receiveAndConvert() throws JmsException

public Object receiveAndConvert(String destinationName)
                                throws JmsException

public Object receiveAndConvert(Destination destination)
                                throws JmsException
```

Each of these methods will use Spring's default message converter (SimpleMessageConverter) or alternatively a custom message converter specified in the messageConverter property on the JmsTemplate bean. Like the receive methods, the receiveAndConvert methods have corresponding receiveSelectedAndConvert methods that accept a message selector as a String argument:

```
public Object receiveSelectedAndConvert(String messageSelector)
                                       throws JmsException

public Object receiveSelectedAndConvert(String destinationName,
                                       String messageSelector)
                                       throws JmsException

public Object receiveSelectedAndConvert(Destination destination,
                                       String messageSelector)
                                       throws JmsException
```

The remaining sections of this chapter cover the details on how to configure and use the various `JmsTemplate` methods to send and receive messages.

Connection Factories and JMS Destinations

The `JmsTemplate` class handles all of the logic for connecting to a JMS provider and accessing the JMS destinations. However, you still need to specify *how* the `JmsTemplate` class establishes the connection to the JMS provider. There are two ways to do this; using JNDI, or using the native connection factories and destination classes supplied by the provider. This section will go through the details of both of these methods, starting with JNDI.

Using JNDI

Spring provides several classes within its messaging framework for accessing JNDI-based connection factories and destinations. The advantage of this approach is that using JNDI further decouples your application from the JMS provider. The Spring classes needed when using JNDI to access the connection factories and JMS destinations are the `JndiTemplate`, `JndiObjectFactoryBean`, `CachingConnectionFactory` (or `SingleConnectionFactory`), `JndiDestinationResolver`, and finally the `JmsTemplate`. The relationship between these Spring classes and the JMS template is shown in Figure 9-4.

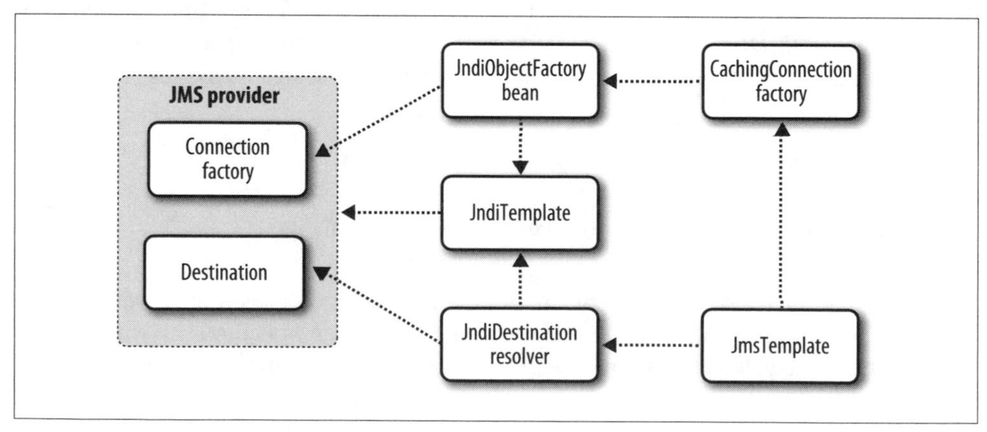

Figure 9-4. JNDI objects and the JmsTemplate

The first thing you need to do is define the `JndiTemplate` bean, which is used to specify the initial context factory, provider URL, and security credentials necessary to make a connection to the JMS provider. It is here that you would specify the TCP address and port for the JMS provider server, which can usually be found in the provider documentation or logs used by the JMS provider. You can refer to Appendix D for more information regarding how to configure ActiveMQ for running the examples in this chapter. The following `JndiTemplate` XML code snippet connects to the JMS provider (in this case ActiveMQ, an open source JMS provider) using the localhost address on port 61616:

```
<bean id="jndiTemplate" class="org.springframework.jndi.JndiTemplate">
    <property name="environment">
       <props>
          <prop key="java.naming.factory.initial">
                org.apache.activemq.jndi.ActiveMQInitialContextFactory</prop>
          <prop key="java.naming.provider.url">tcp://localhost:61616</prop>
          <prop key="java.naming.security.principal">system</prop>
          <prop key="java.naming.security.credentials">manager</prop>
       </props>
    </property>
</bean>
```

Next, you will need to define a bean for the JMS connection factory, which is usually accessed through JNDI. To do this, you would use Spring's `JndiObjectFactoryBean` combined with Spring's `CachingConnectionFactory` (or alternatively, the `SingleConnectionFactory`). The `JndiObjectFactoryBean` takes a reference to the `JndiTemplate` and a `String` argument containing the JNDI name of the connection factory (defined in the JMS provider). The `CachingConnectionFactory`, which is a subclass of the `SingleConnectionFactory`, then takes a reference to the connection factory defined by the `JndiObjectFactoryBean` (in this case `jndiQueueConnectionFactory`) and an integer value indicating the size of the JMS `Session` cache. In the following example, the cache size is set to 1. This value can be increased to support more sessions in the cache:

```
<bean id="jndiQueueConnectionFactory"
      class="org.springframework.jndi.JndiObjectFactoryBean">
    <property name="jndiTemplate" ref="jndiTemplate"/>
    <property name="jndiName" value="QueueCF"/>
</bean>

<bean id="queueConnectionFactory"
      class="org.springframework.jms.connection.CachingConnectionFactory">
   <property name="targetConnectionFactory" ref="jndiQueueConnectionFactory"/>
   <property name="sessionCacheSize" value="1"/>
</bean>
```

After defining the connection factory, you will need to define a `JndiDestinationResolver` bean, which allows you to specify JMS queue and topic destinations by JNDI name rather than as `javax.jms.Destination` reference. The `JmsTemplate` bean will use the `JndiDestinationResolver` to create the JMS destination from the JNDI name. This bean takes a reference to the `JndiTemplate` and has properties for modifying the behavior when resolving JNDI names. The cache property (which defaults to a value of true) will cache resolved destination names. The JndiDestinationResolver will cache dynamic queues and topics by JNDI name, so if using dynamic queues and topics, make sure the names are unique across all queues and topics. The other setting you should be aware of is the fallbackToDynamicDestination boolean flag, which determines whether the JndiDestinationResolver should create a dynamic destination if the JNDI name is not found. If this flag is set to false, it will not create a dynamic destination. If set to true, then a dynamic destination will be created using a name based on the JNDI destination name. The default value for this flag is false. In the following example, the JndiDestinationResolver bean is defined using the JndiTemplate previously created using caching and configured not to create dynamic destinations:

```
<bean id="destinationResolver"
      class="org.springframework.jms.support.destination.JndiDestinationResolver">
   <property name="jndiTemplate" ref="jndiTemplate"/>
   <property name="cache" value="true"/>
   <property name="fallbackToDynamicDestination" value="false"/>
</bean>
```

The last step in the configuration process is to create the JmsTemplate bean. When using JNDI, the basic properties for the JmsTemplate include the connection factory, the destination resolver, a pubSubDomain property indicating whether you are using queues or topics for this JMS template, and optionally a default destination for the template. As illustrated in the following code, the JmsTemplate binds together the connection factory, destination resolver, and a default destination name (in this case queue1). When using JMS 1.1, you do not need to worry about the pubSubDomain property unless you are using dynamic destination creation:

```
<bean id="jmsTemplate" class="org.springframework.jms.core.JmsTemplate">
   <property name="connectionFactory" ref="queueConnectionFactory"/>
   <property name="destinationResolver" ref="destinationResolver"/>
   <property name="defaultDestinationName" value="queue1"/>
   <property name="pubSubDomain" value="false"/>
</bean>
```

The following code shows all of the Spring beans necessary for connecting to the JMS provider, defining the connection factory, and specifying the JMS destinations when using JNDI. This configuration will be used for the synchronous sender and receiver examples that follow:

```
...

<bean id="jndiTemplate" class="org.springframework.jndi.JndiTemplate">
   <property name="environment">
      <props>
         <prop key="java.naming.factory.initial">
                    org.apache.activemq.jndi.ActiveMQInitialContextFactory</prop>
         <prop key="java.naming.provider.url">tcp://localhost:61616</prop>
         <prop key="java.naming.security.principal">system</prop>
         <prop key="java.naming.security.credentials">manager</prop>
      </props>
```

```
    </property>
</bean>

<bean id="jndiQueueConnectionFactory"
      class="org.springframework.jndi.JndiObjectFactoryBean">
   <property name="jndiTemplate" ref="jndiTemplate"/>
   <property name="jndiName" value="QueueCF"/>
</bean>

<bean id="queueConnectionFactory"
      class="org.springframework.jms.connection.CachingConnectionFactory">
   <property name="targetConnectionFactory" ref="jndiQueueConnectionFactory"/>
   <property name="sessionCacheSize" value="1"/>
</bean>

<bean id="destinationResolver"
      class="org.springframework.jms.support.destination.JndiDestinationResolver">
   <property name="jndiTemplate" ref="jndiTemplate"/>
   <property name="cache" value="true"/>
   <property name="fallbackToDynamicDestination" value="false"/>
</bean>

<bean id="jmsTemplate" class="org.springframework.jms.core.JmsTemplate">
   <property name="connectionFactory" ref="queueConnectionFactory"/>
   <property name="destinationResolver" ref="destinationResolver"/>
   <property name="defaultDestinationName" value="queue1"/>
   <property name="pubSubDomain" value="false"/>
</bean>

...
```

Using Native Classes

As an alternative to using JNDI, you can define the connection factory and JMS destination beans using native JMS provider classes. This is a useful alternative if the JMS provider does not use JNDI or does not support JNDI connection factories or destinations. Using this approach, you do not specify a `JndiTemplate` or `JndiDestinationResolver`, nor do you use the `JndiObjectFactoryBean`. Instead, the connection factory and JMS destinations are defined using the JMS provider classes directly.

Each native JMS provider class will have a different set of properties used to specify the URL, port, and other information for connecting to the JMS provider. For example, IBM WebSphere MQ v6 uses the `MQQueueConnectionFactory` class located in the `com.ibm.mq.jms` package, which takes a transport type, queue manager name, hostname, port, and channel. Note that in this example, we are leveraging the `FieldRetrievingFactoryBean` to convert the static field reference to the actual value rather than specifying the integer value for the `transportType` in the `MQQueueConnectionFactory`:

```
<bean id="transport"
      class="org.springframework.beans.factory.config.FieldRetrievingFactoryBean">
   <property name="staticField">
         <value>com.ibm.mq.jms.JMSC.MQJMS_TP_CLIENT_MQ_TCPIP</value>
```

```
    </property>
</bean>

<bean id="queueConnectionFactory" class="com.ibm.mq.jms.MQQueueConnectionFactory">
    <property name="transportType" ref="transport" />
    <property name="queueManager" value="QM1" />
    <property name="hostName" value="localhost" />
    <property name="port" value="1415" />
    <property name="channel" value="SYSTEM.DEF.SVRCONN" />
</bean>
```

To illustrate how the native connection factories differ, consider the following example, which uses the native classes in ActiveMQ, a robust open source JMS provider:

```
<bean id="queueConnectionFactory" class="org.activemq.ActiveMQConnectionFactory">
    <property name="brokerURL" value="tcp://localhost:61616"/>
</bean>
```

Notice that ActiveMQ only requires the `brokerURL` property needs to be defined, whereas IBM WebSphere MQ requires the transport type, queue manager name, hostname, port, and channel. Using the native JMS provider classes binds your code specifically to that provider, whereas using JNDI does not.

For those connection factories that require a secure connection (such as IBM WebSphere MQ), you can wrap the connection factory bean in Spring's `UserCredentialsConnectionFactoryAdapter` bean. This bean takes the connection factory defined previously as an argument, along with the username and password. Note that if you are not connecting to IBM WebSphere MQ using security credentials, you still need to define the secure connection factory passing in a space in the username and password property values.

```
<bean id="secureQueueConnectionFactory"
      class=
        "org.springframework.jms.connection.UserCredentialsConnectionFactoryAdapter">
    <property name="targetConnectionFactory" ref="queueConnectionFactory"/>
    <property name="username" value="admin"/>
    <property name="password" value="pwd"/>
</bean>
```

Defining JMS destinations is simply a matter of creating a bean with the native destination class, usually providing a constructor argument containing the name of the queue or topic. For example, in IBM WebSphere MQ you would create a JMS destination bean using the `com.ibm.mq.jms.MQQueue` native class, as follows:

```
<bean id="queueDest1" class="com.ibm.mq.jms.MQQueue">
    <constructor-arg value="queue1" />
</bean>
```

Once all of the connection factories and JMS destinations have been defined, you can create the `JmsTemplate` bean using the native JMS connection factory and JMS destination beans:

```
<bean id="jmsTemplate" class="org.springframework.jms.core.JmsTemplate">
    <property name="connectionFactory" ref="secureQueueConnectionFactory"/>
```

```
    <property name="defaultDestination" ref="queueDest1"/>
    <property name="pubSubDomain" value="false"/>
</bean>
```

The complete listing for configuring Spring to use native classes (in this example, IBM WebSphere MQ) is as follows:

```
<bean id="transport"
      class="org.springframework.beans.factory.config.FieldRetrievingFactoryBean">
    <property name="staticField">
        <value>com.ibm.mq.jms.JMSC.MQJMS_TP_CLIENT_MQ_TCPIP</value>
    </property>
</bean>

<bean id="queueConnectionFactory" class="com.ibm.mq.jms.MQQueueConnectionFactory">
    <property name="transportType" ref="transport" />
    <property name="queueManager" value="QM1" />
    <property name="hostName" value="localhost" />
    <property name="port" value="1415" />
    <property name="channel" value="SYSTEM.DEF.SVRCONN" />
</bean>

<bean id="secureQueueConnectionFactory"
      class=
        "org.springframework.jms.connection.UserCredentialsConnectionFactoryAdapter">
    <property name="targetConnectionFactory" ref="queueConnectionFactory"/>
    <property name="username" value="admin"/>
    <property name="password" value="pwd"/>
</bean>

<bean id="queueDest1" class="com.ibm.mq.jms.MQQueue">
    <constructor-arg value="queue1" />
</bean>

<bean id="jmsTemplate" class="org.springframework.jms.core.JmsTemplate">
    <property name="connectionFactory" ref="secureQueueConnectionFactory"/>
    <property name="defaultDestination" ref="queueDest1"/>
    <property name="pubSubDomain" value="false"/>
</bean>
```

Sending Messages

There are two ways to send a message when using Spring: you can create a standard JMS `Message` object (e.g., `TextMessage`) and send that message using the `send` method on the `JmsTemplate`, or you can create a Java object (e.g., `String`) and send it using the `convertAndSend` method without having to first create a JMS `Message` object. Both of these techniques can use either a default destination (specified in the `defaultDestination` or `defaultDestinationName` properties on the `JmsTemplate`) or a specific destination specified directly in the `send` or `convertAndSend` methods.

When specifying a default destination on the JmsTemplate, you can use either the defaultDestination property or the defaultDestinationName property. The defaultDestination property takes a reference to a JMS destination. This can be a javax.jms.Destination, javax.jms.Queue, javax.jms.Topic, or a reference bean created from one of the specific destination classes provided by a JMS provider (for example, com.ibm.mq.jms.MQQueue). The defaultDestinationName property takes a String argument containing the JNDI name of the queue or topic. Note that when using the defaultDestinationName, you must also have a destinationResolver bean defined so the JNDI name can be resolved.

In the following two subsections, we will show how to use the send and convertAndSend methods using a default destination. In the third subsection, we will show how to specify a JMS destination in the send methods using either a JNDI name or a javax.jms.Destination object. This is useful when you have multiple queues or topics serviced by a single JmsTemplate or have a variable queue or topic.

Using the send Method

To use the send method in the JmsTemplate, you must first create the JMS Message object you want to send using Spring's MessageCreator class. The MessageCreator class is used to construct the JMS Message object, which is then used by the JmsTemplate send method. The MessageCreator class can be created as an anonymous inner class as part of the body of the send method or instantiated outside of the context of the send method. When creating the MessageCreator object, you need override the createMessage method, which returns a JMS Message object. It is within the createMessage method that you write custom JMS code to create the JMS Message object you wish to send. The following example shows the use of the MessageCreator to create a simple TextMessage:

```
public class SimpleJMSSender {

   public static void main(String[] args) {
     try {
       ApplicationContext ctx =
           new ClassPathXmlApplicationContext("app-context.xml");
       JmsTemplate jmsTemplate =
           (JmsTemplate)ctx.getBean("jmsTemplate");

       MessageCreator mc = new MessageCreator() {
         public Message createMessage(Session session) throws JMSException {
            TextMessage msg = session.createTextMessage();
            msg.setJMSPriority(9);
            msg.setText("This is easy!");
            return msg;
         }
       };

       jmsTemplate.send(mc);
     } catch (Exception ex) {
```

```
            ex.printStackTrace();
        }
    }
}
```

Notice that the `createMessage` method contains the JMS `Session` object, which allows you to create the JMS `Message` object. In this example, the JMS priority header property is set to a value of 9 (high priority), illustrating that header and application properties can be set in this method as well. The `MessageCreator` object is passed into the `send` method, which invokes the `createMessage` method to get the JMS message object to send.

An alternative technique that yields the same results is to instantiate the `MessageCreator` object as an anonymous inner class within the context of the `send` method:

```
public class SimpleJMSSender {

    public static void main(String[] args) {
        try {
            ApplicationContext ctx =
                new ClassPathXmlApplicationContext("app-context.xml");
            JmsTemplate jmsTemplate =
                (JmsTemplate)ctx.getBean("jmsTemplate");

            jmsTemplate.send(new MessageCreator() {
                public Message createMessage(Session session) throws JMSException {
                    TextMessage msg = session.createTextMessage();
                    msg.setJMSPriority(9);
                    msg.setText("This is easy!");
                    return msg;
                }
            });
        } catch (Exception ex) {
            ex.printStackTrace();
        }
    }
}
```

Notice that when using the `JmsTemplate` you do not need to connect to the JMS provider or create a `QueueSession` and `QueueSender` object. All of this infrastructure-related messaging logic is handled by the `JmsTemplate`, making it easier to send messages using the Spring Framework.

Using the convertAndSend Method

One of the implications of using the `send` method is that you still have to use the JMS API to create the JMS `Message` object. Spring offers another approach that even further simplifies the task of sending a message: automatic message conversion. The `convertAndSend` method is used to send a message without having to create the JMS `Message` object. This method will automatically convert a Java object into its corresponding JMS `Message` type using a *message converter*. Spring's default message converter (SimpleMessageConverter) will convert a Java object into an ObjectMessage, a String object into a TextMessage, a byte[] array into a BytesMessage, and a java.util.Map object into a MapMessage. You can define other custom conversions by writing your own message converter and wiring it to the JmsTemplate using the messageConverter property.

Using the example from the previous section, you can send a simple TextMessage using the default message converter as follows:

```
public class SimpleJMSSender {

   public static void main(String[] args) {
     try {
       ApplicationContext ctx =
           new ClassPathXmlApplicationContext("app-context.xml");
       JmsTemplate jmsTemplate =
           (JmsTemplate)ctx.getBean("jmsTemplate");

       jmsTemplate.convertAndsend("This is even easier!");

     } catch (Exception ex) {
         ex.printStackTrace();
     }
   }
}
```

Notice in this example you do not need to create a TextMessage object. It is automatically created for you using the String value passed into the convertAndSend method. Using this technique, you can reduce the Java code used to send a simple message to one line of code.

Since the actual message object is not accessible when sending an object, how do you set the header properties and application properties on the message? Easy. To set message header and application properties when using the convertAndSend method, you can create a Spring MessagePostProcessor object, which gives you access to the JMS Message object created by the convertAndSend method. The MessagePostProcessor object can be created as an inner class or as an anonymous inner class within the convertAndSend method.

To illustrate the use of the MessagePostProcessor object, consider the following example where you want to set the priority of the message via the JMSPriority header property. Here you can simply create a MessagePostProcessor object and override the postProcessMessage method to set the message priority:

```
public class SimpleJMSSender {

   public static void main(String[] args) {
     try {
       ApplicationContext ctx =
           new ClassPathXmlApplicationContext("app-context.xml");
       JmsTemplate jmsTemplate =
           (JmsTemplate)ctx.getBean("jmsTemplate");
```

```
        MessagePostProcessor postProcessor = new MessagePostProcessor() {
            public Message postProcessMessage(Message message)
                   throws JMSException {
               message.setJMSPriority(9);
               return message;
            }
        };

        jmsTemplate.convertAndsend((Object)"This is even easier!", postProcessor);

      } catch (Exception ex) {
         ex.printStackTrace();
      }
   }
}
```

As illustrated in the example just shown, when using the `MessagePostProcessor` in conjunction with a `String` message type with a default destination, you must cast the `String` message body to an `Object` to avoid ambiguity with the overloaded method signatures on the `convertAndSend` method. Within the `postProcessMessage` method you can also do other messaging related tasks, such as set application properties on the message. This technique provides the same capabilities as you have when creating the JMS `Message` object directly.

Using a Nondefault JMS Destination

While having a default destination specified in the `JmsTemplate` is handy, there are situations in which you may have a common `JmsTemplate` that is shared by multiple sender classes, each using a different JMS destination. In this case, you need to specify the destination in the `send` method itself. There are two ways of doing this: use the JNDI destination name or define a JMS destination object and use that reference in the `send` method.

To use a JNDI destination name, the `JmsTemplate` needs to have a reference to a `JNDI DestinationResolver`, which is specified in the `destinationResolver` property. Using the JNDI name is simply a matter of setting the `String` value of the JNDI destination name (in this case `queue1`) in the `send` method as illustrated here:

```
public class SimpleJMSSender {

   public static void main(String[] args) {
     try {
       ApplicationContext ctx =
           new ClassPathXmlApplicationContext("app-context.xml");
       JmsTemplate jmsTemplate =
           (JmsTemplate)ctx.getBean("jmsTemplate");

      jmsTemplate.send("queue1", new MessageCreator() {
         public Message createMessage(Session session) throws JMSException {
            TextMessage msg = session.createTextMessage();
```

```
            msg.setJMSPriority(9);
            msg.setText("This is easy!");
            return msg;
         }
      });
     } catch (Exception ex) {
        ex.printStackTrace();
     }
   }
}
```

In the example just shown, the JMS destination with the JNDI name `queue1` will be used when sending messages. Rather than hardcoding this value, you will obviously want to either pass this value in as an argument or specify the destination JNDI name as a property for this class within the Spring application context XML file (assuming it is defined as a Spring-managed bean). The `JmsTemplate` will use the `JNDIDestinationResolver` to resolve the JNDI name within the `send` method.

The other alternative is to use a `javax.jms.Destination` object as the argument for the JMS destination in the `send` method. To do this you would first create a Spring bean containing the JNDI destination using the `JndiObjectFactoryBean` from Spring:

```
...
<bean id="queue1Dest"
   class="org.springframework.jndi.JndiObjectFactoryBean">
   <property name="jndiTemplate" ref="jndiTemplate"/>
   <property name="jndiName" value="queue1"/>
</bean>
```

Now, in the code, you can either get the `queue1Dest` bean from the Spring application context directly via the `getBean` method or add a property to the application sender bean if it is a Spring managed bean:

```
public class SimpleJMSSender {

   public static void main(String[] args) {
     try {
       ApplicationContext ctx =
           new ClassPathXmlApplicationContext("app-context.xml");
       JmsTemplate jmsTemplate =
           (JmsTemplate)ctx.getBean("jmsTemplate");
       Destination queue =
           (Destination)ctx.getBean("queue1Dest");

      jmsTemplate.send(queue, new MessageCreator() {
         public Message createMessage(Session session) throws JMSException {
            TextMessage msg = session.createTextMessage();
            msg.setJMSPriority(9);
            msg.setText("This is easy!");
            return msg;
         }
      });
     } catch (Exception ex) {
        ex.printStackTrace();
```

```
            }
        }
    }
```

One of the differences between using the JNDI name versus the JMS destination is when the JNDI name is resolved. When passing the JNDI destination name into the `send` method, the name is not resolved until the `send` method is ready to send the message. When using a JMS destination reference, the name is resolved when the JMS destination bean (e.g., `queueDest1`) during the initial loading of the Spring application context. Therefore, when defining a JMS destination bean object, errors in the JNDI name will be realized during application startup rather than when the queue is actually used. While using the JNDI destination name approach requires less configuration, it does require additional testing to discover possible JNDI name resolution errors.

Receiving Messages Synchronously

Receiving messages synchronously works much in the same way that sending a message works in that they both use the `JmsTemplate` class. When receiving messages synchronously the application receiver class blocks until a message is received from the queue or topic. Like the `send` methods on the `JmsTemplate`, there are two forms of `receive` methods; `receive` and `receiveAndConvert`. The `receive` method returns a JMS `Message` object, whereas the `receiveAndConvert` method returns a Java `Object` corresponding to the type of JMS `Message` object received. Like the `send` method, you can use a default destination or specify the JNDI name or JMS destination directly in the `receive` method itself.

The receiver timeout value allows you to control the amount of time the application receiver should wait before receiving a message. The default value is 0, indicating that the receiver should block and wait forever until a message is received. The receiver timeout value can be specified through the `receiveTimeout` property on the `JmsTemplate` bean. This property takes a `long` value containing of the receive timeout in milliseconds. For example, receivers using the `JmsTemplate` defined here will stop waiting for messages after 30 seconds:

```
<bean id="jmsTemplate" class="org.springframework.jms.core.JmsTemplate">
   <property name="connectionFactory" ref="queueConnectionFactory"/>
   <property name="destinationResolver" ref="destinationResolver"/>
   <property name="defaultDestinationName" value="queue1"/>
   <property name="receiveTimeout" value="30000"/>
   <property name="pubSubDomain" value="false"/>
</bean>
```

Like the JMS API, if the `receive` method times out based on the `receiveTimeout` property, the `receive` method will return a null. You should always account for this in your code when using a timeout value. To receive a JMS `Message` object from a JMS destination you would use the `receive` method on the `JmsTemplate` class. In the following code example, the receive method is used to block and wait for a JMS TextMessage object received from a default destination:

```
public class SimpleJMSReceiver {

   public static void main(String[] args) {
     try {
       ApplicationContext ctx =
           new ClassPathXmlApplicationContext("app-context.xml");
       JmsTemplate jmsTemplate =
           (JmsTemplate)ctx.getBean("jmsTemplate");

       Message msg = jmsTemplate.receive();
       if (msg instanceof TextMessage) {
          System.out.println(((TextMessage)msg).getText());
       } else {
          throw new IllegalStateException("Message type not supported");
       }
     } catch (Exception ex) {
        ex.printStackTrace();
     }
   }
}
```

Notice that the receive method on the JmsTemplate returns a standard JMS Message object. Once received, you can process the message just as you would when using the JMS API.

The other alternative for receiving messages synchronously is to have Spring automatically convert the incoming JMS Message object into a corresponding Java object, eliminating the need to use the JMS API altogether. Using the receiveAndConvert method with Spring's default message converter (SimpleMessageConverter), a TextMessage will be converted to a String object, a BytesMessage converted to a byte[] array, a MapMessage converted to a java.util.Map object, and an ObjectMessage converted to a serialized Java Object. The following example illustrates the use of the receiveAndConvert method for receiving a TextMessage from the queue:

```
public class SimpleJMSReceiver {

   public static void main(String[] args) {
     try {
       ApplicationContext ctx =
           new ClassPathXmlApplicationContext("app-context.xml");
       JmsTemplate jmsTemplate =
           (JmsTemplate)ctx.getBean("jmsTemplate");

       Object msg = jmsTemplate.receiveAndConvert();
       if (msg instanceof String) {
          System.out.println(msg);
       } else {
          throw new IllegalStateException("Message type not supported");
       }
     } catch (Exception ex) {
        ex.printStackTrace();
```

```
        }
      }
    }
```

Notice in this example that the `SimpleJMSReceiver` class does not reference the JMS API at all. You can also define your own custom message converter to convert other types of messages or enhance the incoming message with additional data or properties.

There may be times when you want to be more selective about the messages you receive from a queue or topic. You can apply a *message selector* to filter only those messages you are interested in receiving (see Chapter 6 for details on message selectors). Spring provides additional `receive` methods for specifying a message selector for receiving messages synchronously: `receiveSelected`, which corresponds to the `receive` method, and `receiveSelectedAndConvert`, which corresponds to the `receiveAndConvert`. In addition to the arguments in the corresponding `receive` and `receiveAndConvert` methods, these two methods also take an additional `String` argument containing the message selector to apply for this receiver. For example, to filter out normal priority messages and receive only high priority messages (those with a JMS priority greater than 4), you would use the `receiveSelected` method, passing in the appropriate message selector:

```
public class SimpleJMSReceiver {

   public static void main(String[] args) {
     try {
       ApplicationContext ctx =
           new ClassPathXmlApplicationContext("app-context.xml");
       JmsTemplate jmsTemplate =
           (JmsTemplate)ctx.getBean("jmsTemplate");

       Message msg = jmsTemplate.receiveSelected("JMSPriority > 4");
       if (msg instanceof TextMessage) {
          System.out.println(((TextMessage)msg).getText());
       } else {
          throw new IllegalStateException("Message type not supported");
       }
     } catch (Exception ex) {
        ex.printStackTrace();
     }
   }
}
```

You would normally receive messages synchronously when using the request/reply model, where you send a message and then block and wait for a response. For example, you may send a message to create an order for a customer, but you need to wait for the response to confirm the order creation and have the order number returned to you before continuing. For those times when you want to create a nonblocking listener, you would use asynchronous processing using MDPs. MDPs are described in the next section.

Message-Driven POJOs

Receiving messages asynchronously means you have a nonblocking process that is listening for messages on a particular queue or topic. This technique is a form of event-driven processing where the presence of a message triggers an action on a message listener. Message-driven beans (discussed in Chapter 8) are Java EE's answer to asynchronous receivers. The Spring Framework also supports asynchronous receivers through MDPs.

There are three different ways of configuring asynchronous message listeners in Spring: implementing the `javax.jms.MessageListener` interface, implementing Spring's `SessionAwareMessageListener`, and finally, wrapping a standard POJO in Spring's `MessageListenerAdapter` class. These three methods vary in terms of how the message listener class is structured. All three of these methods use a *message listener container*, which is analogous to the `JmsTemplate` class described in the previous sections. The following sections will describe the details of the message listener container and each of the three message-driven bean techniques.

The Spring Message Listener Container

Message-driven POJOs are created within the context of a message listener container. The message listener container binds the connection factory, JMS destination, JNDI destination resolver (if using JNDI), and the message listener bean. Spring provides two message listener containers: the `DefaultMessageListenerContainer` and the `SimpleMessageListenerContainer`. While both of these message listener containers allow you to specify the number of concurrent listener threads, only the `DefaultMessageListenerContainer` has the ability to dynamically adjust the number of listener threads during runtime. In addition, the `DefaultMessageListenerContainer` allows integration with XA transactions, whereas the `SimpleMessageListenerContainer` does not. For simple messaging applications that use a local transaction manager and do not require thread, session, or connection tuning based on varying load conditions, use the `SimpleMessageListenerContainer`. For messaging applications using an external transaction manager or XA transactions that may require tuning, use the `DefaultMessageListenerContainer`.

The configuration of the message listener container is similar to the `JmsTemplate`; you can use JNDI to access the connection factory and JMS destinations or use the JMS provider's native connection factory and JMS destination classes directly. The following example shows the configuration necessary to setup a message-driven POJO (`SimpleJmsReceiver`) using the `DefaultMessageListenerContainer` with the ActiveMQ JMS provider:

```
...

<bean id="jndiTemplate" class="org.springframework.jndi.JndiTemplate">
  <property name="environment">
```

```
      <props>
        <prop key="java.naming.factory.initial">
              org.apache.activemq.jndi.ActiveMQInitialContextFactory</prop>
        <prop key="java.naming.provider.url">tcp://localhost:61616</prop>
        <prop key="java.naming.security.principal">system</prop>
        <prop key="java.naming.security.credentials">manager</prop>
      </props>
    </property>
  </bean>

  <bean id="jndiQueueConnectionFactory"
        class="org.springframework.jndi.JndiObjectFactoryBean">
    <property name="jndiTemplate" ref="jndiTemplate"/>
    <property name="jndiName" value="QueueCF"/>
  </bean>

  <bean id="queueConnectionFactory"
        class="org.springframework.jms.connection.CachingConnectionFactory">
     <property name="targetConnectionFactory" ref="jndiQueueConnectionFactory"/>
     <property name="sessionCacheSize" value="1"/>
  </bean>

  <bean id="destinationResolver"
        class="org.springframework.jms.support.destination.JndiDestinationResolver">
     <property name="jndiTemplate" ref="jndiTemplate"/>
     <property name="cache" value="true"/>
     <property name="fallbackToDynamicDestination" value="false"/>
  </bean>

  <bean id="messageListener" class="SimpleJMSReceiver"/>

  <bean id="jmsContainer"
        class="org.springframework.jms.listener.DefaultMessageListenerContainer">
    <property name="connectionFactory" ref="queueConnectionFactory"/>
    <property name="destinationResolver" ref="destinationResolver"/>
    <property name="concurrentConsumers" value="3" />
    <property name="destinationName" value="queue1"/>
    <property name="messageListener" ref="messageListener" />
  </bean>

  ...
```

Notice in this configuration how the `messageListener` bean is wired to the `DefaultMessageListenerContainer` through the `jmsContainer` bean, along with the destination name, connection factory, and JNDI destination resolver. In this example, three concurrent listener threads will be started, as indicated by the `concurrentConsumers` property. This configuration will be used for the rest of this section to illustrate the three ways to implement a message listener.

MDP Option 1: Using the MessageListener Interface

The simplest form of a message-driven POJO is an asynchronous receiver that implements the `javax.jms.MessageListener` interface. This is similar to the EJB3 message-driven beans described in Chapter 8. As illustrated in Figure 9-5, the POJO receiver (e.g., `SimpleJMSReceiver`) implements the `javax.jms.MessageListener` interface, which is then wired to the `DefaultMessageListenerContainer`. The `DefaultMessageListenerContainer` is then wired to the `CachingConnectionFactory` and the `JNDIDestinationResolver` (assuming you are using JNDI).

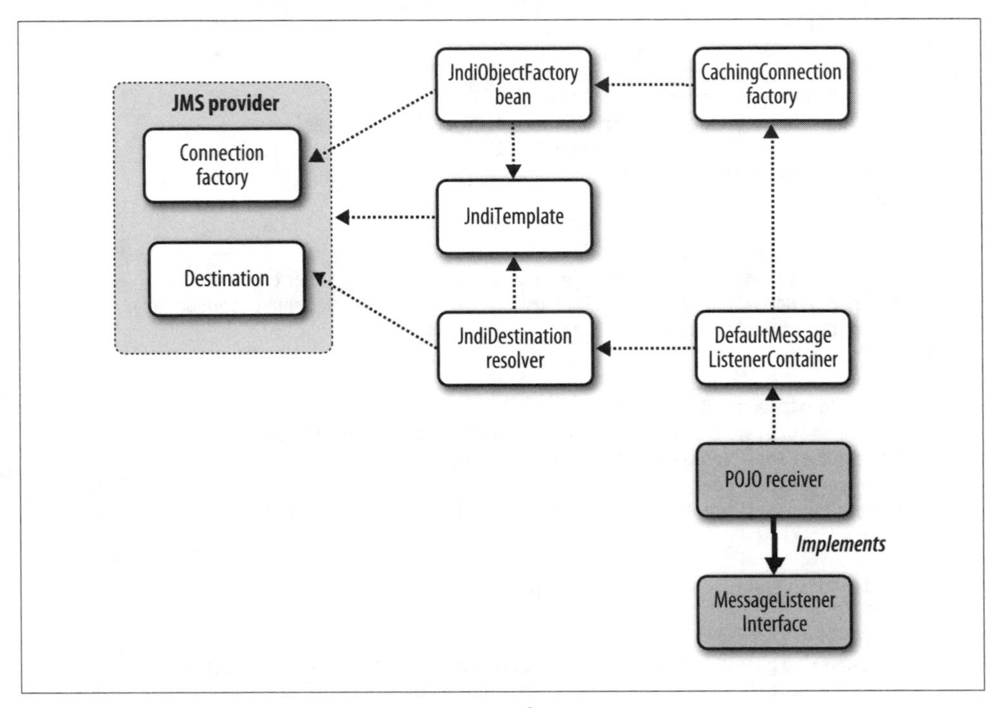

Figure 9-5. Implementing the MessageListener interface

When implementing the `MessageListener` interface, you must override the `onMessage` method in your message listener class. There are no changes to the prior XML configuration when using this technique. In the following code example, the `SimpleJMSReceiver` message listener implements the `javax.jms.MessageListener` interface and overrides the `onMessage` method, which in turn consumes a `TextMessage` object:

```
public class SimpleJMSReceiver implements MessageListener {

  public void onMessage(Message message) {
    try {
      if (message instanceof TextMessage) {
         System.out.println(((TextMessage)message).getText());
      } else {
         throw new IllegalStateException("Message Type Not Supported");
      }
    } catch (JMSException e) {
      e.printStackTrace();
    }
```

```
    }
}
```

MDP Option 2: Using the SessionAwareMessageListener Interface

The Spring Framework provides an extension to the `javax.jms.MessageListener` interface called the `SessionAwareMessageListener`. Like the `javax.jms.MessageListener` interface, the `SessionAwareMessageListener` interface contains an `onMessage` method that must be overridden by the listener class. However, unlike the `javax.jms.MessageListener` interface, in addition to the `Message` object, the `SessionAwareMessageListener` interface also provides access to the JMS `Session`:

```
void onMessage(Message message,
               Session session)
               throws JMSException
```

Figure 9-6 illustrates the use of the `SessionAwareMessageListener` technique. From a configuration standpoint, this technique is the same as when using the `javax.jms.MessageListener` interface.

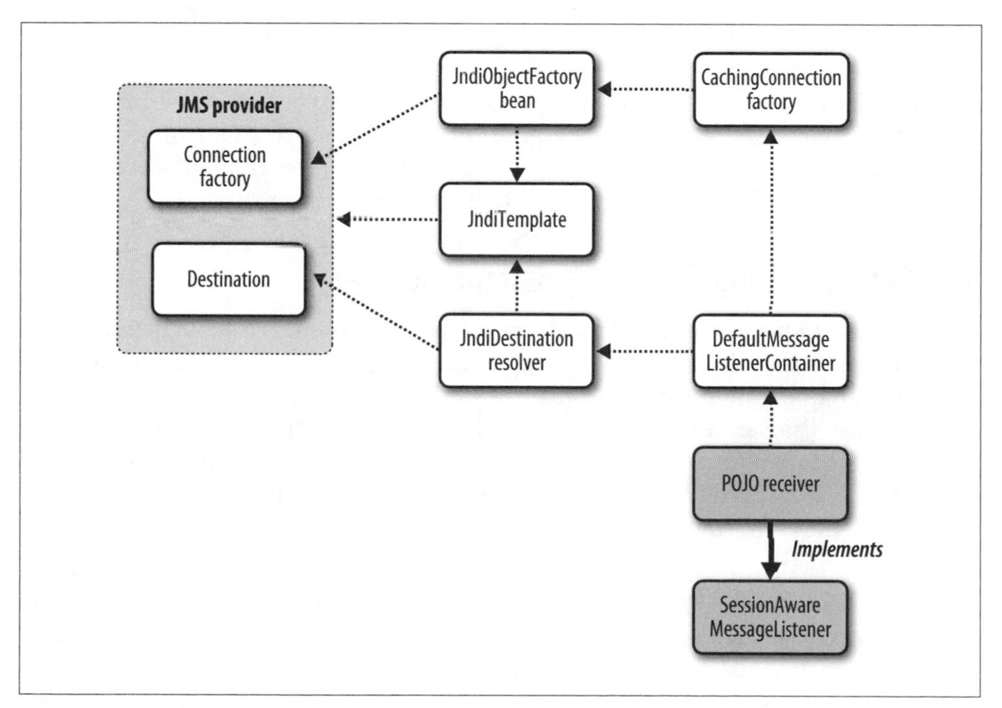

Figure 9-6. Implementing the SessionAwareMessageListener interface

The `SessionAwareMessageListener` is useful when you need access to the JMS `Session` object in the asynchronous message listener. One common use case for this is when you need to send a response message back to the sender. Another use case is when you need to transact the Session. Consider the following simple example where the SimpleJMSReceiver class returns a message back to the sender on the destination specified in the JMSReplyTo header property indicating that it processed the message:

```
public class SimpleJMSReceiver implements SessionAwareMessageListener {

   public void onMessage(Message message, Session session) throws JMSException {
      if (message instanceof TextMessage) {
         String text = ((TextMessage)message).getText();
         System.out.println(text);

         //send the response
         MessageProducer sender =
            session.createProducer(message.getJMSReplyTo());
         TextMessage msg = session.createTextMessage();
         msg.setJMSCorrelationID(message.getJMSMessageID());
         msg.setText("Message " + message.getJMSMessageID()
                              + " received");
         sender.send(msg);
      } else {
         throw new IllegalStateException("Message type not supported");
      }
   }
}
```

Notice that you do not have to obtain a Session object yourself or clean up after you sent the message; Spring handles all of this in the JmsTemplate. Also notice that the onMessage method of Spring's SessionAwareMessageListener interface throws a JMSException, whereas the javax.jms.MessageListener interface does not.

MDP Option 3: Using the MessageListenerAdapter

The third technique for creating asynchronous message listeners is to wrap a standard POJO receiver class in a Spring MessageListenerAdapter object. What makes this technique unique from the others is that the POJO receiver class does not implement any message listener interface, nor does it need to include any reference to a javax.jms.Message object. As illustrated in Figure 9-7, the POJO receiver class is wired to Spring's MessageListenerAdapter, which is then referenced by the messageListener property on the DefaultMessageListenerContainer.

There are several ways to structure your POJO receiver methods when using the MessageListenerAdapter. You can use the default message handling methods used by the MessageListenerAdapter or designate a custom method on the listener class to be the listener method. When using the latter, you can also specify whether to use a message converter to convert the incoming message into the corresponding Java object type or consume a JMS Message object directly. Both of these options will be explored in the following sections.

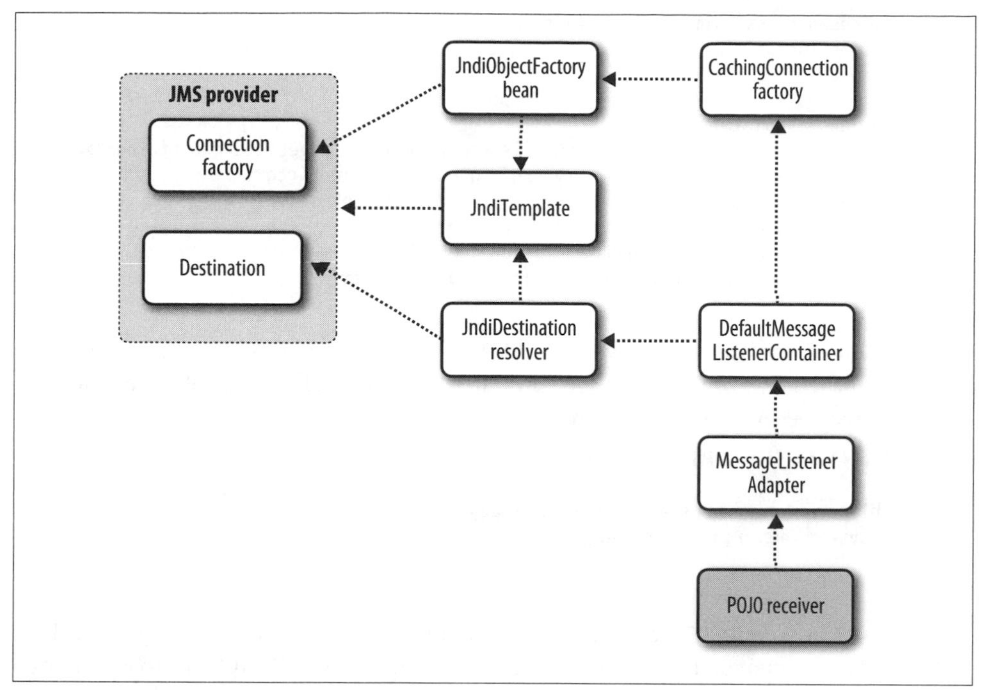

Figure 9-7. Using a MessageListenerContainer bean

Default message handler method

By default, the `MessageListenerAdapter` will look for a `handleMessage` method in your POJO corresponding to the type of JMS message being received. The following listing contains the `handleMessage` method signatures when using the automatic message conversion:

```
//receive a converted TextMessage
public void handleMessage(String message) {...}

//receive a converted BytesMessage
public void handleMessage(byte[ ] message) {...}

//receive a converted MapMessage
public void handleMessage(Map message) {...}

//receive a converted ObjectMessage
public void handleMessage(Object message) {...}
```

To use the default message listener handler methods, all you need to do is wire the message-driven POJO (e.g., `SimpleJMSReceiver`) to the `MessageListenerAdapter` bean through the `constructor-arg` property (or `delegate` property):

```
<bean id="messageListener"
      class="org.springframework.jms.listener.adapter.MessageListenerAdapter">
   <constructor-arg>
```

```
        <bean class="SimpleJMSReceiver"/>
      </constructor-arg>
    </bean>

    <bean id="jmsContainer"
        class="org.springframework.jms.listener.DefaultMessageListenerContainer">
      <property name="connectionFactory" ref="queueConnectionFactory"/>
      <property name="destinationResolver" ref="destinationResolver"/>
      <property name="concurrentConsumers" value="3" />
      <property name="destinationName" value="queue1"/>
      <property name="messageListener" ref="messageListener" />
    </bean>
```

When defining the message-driven POJO, you would simply define a `handleMessage` method for each type of JMS message you want to consume. For example, the following code will receive JMS `TextMessage` messages:

```
public class SimpleJMSReceiver {

  public void handleMessage(String message) {
     System.out.println(message);
  }
}
```

Notice that the `SimpleJMSReceiver` class does not have a reference to any JMS API interface; as a matter of fact, in this example the message-driven POJO has no knowledge whatsoever that it is even being used within the context of messaging. All of the messaging infrastructure is handled by the `MessageListenerAdapter` and the `DefaultMessageListenerContainer`. All you have to do is write the necessary POJO `handleMessage` methods that correspond to the type of JMS message being received.

What happens if you are not sure of the JMS `Message` type you will be receiving, or what if you have the possibility of receiving either a `TextMessage` or `MapMessage` to process? In prior examples you could check the instance of the JMS `Message` object and direct processing based on the `Message` type:

```
public class SimpleJMSReceiver implements MessageListener {

  public void onMessage(Message message) {
    try {
      if (message instanceof TextMessage) {
         //process message text...
      } else if (message instanceof MapMessage) {
         //process map message...
      } else {
         throw new IllegalStateException("Message Type Not Supported");
      }
    } catch (JMSException e) {
       e.printStackTrace();
    }
  }
}
```

However, the argument in the default handleMessage method is already "cast" to the message type it is supposed to receive. Spring handles this in several ways: if the corresponding default handleMessage method for a particular message type is not found in the class, the MessageListenerAdapter will throw (and then absorb) a NoSuchMethodException indicating that the corresponding message handler method for that JMS message type was not found. Therefore, you must code a handleMessage method for each type of message you expect to receive. For example, if you are expecting to receive TextMessage and MapMessage message types, then you only need to code the handleMessage methods containing a String and Map argument:

```
public class SimpleJMSReceiver {

  public void handleMessage(String message) {
     //process String message body
  }

  public void handleMessage(Map message) {
     //process Map message body
  }
}
```

One issue with the message conversion just described is that only the message payload is passed into the message handler method. Therefore, you do not have access to any of the message header properties or message application properties that may have been set on the message. For example, the sender may have used the application properties section of the message to pass additional metadata about the message (e.g., security credentials) or you may need access to certain message header properties such as the JMSReplyTo property or the JMSMessageID property. In these cases, you can tell the DefaultMessageListenerContainer that you do not wish to have the message payload automatically converted. You can easily do this by setting the messageConverter property value to null in the MessageListenerAdapter bean, which tells it to look for a handleMessage method with a corresponding JMS Message object argument rather than the corresponding Java object type:

```
<bean id="messageListener"
      class="org.springframework.jms.listener.adapter.MessageListenerAdapter">
   <constructor-arg>
      <bean class="SimpleJMSReceiver"/>
   </constructor-arg>
   <property name="messageConverter"><null/></property>
</bean>
```

When disabling the message conversion feature, the MessageListenerAdapter will, by default, look for one of the following handleMessage methods:

```
//receive a JMS TextMessage
public void handleMessage(TextMessage message) {...}

//receive a JMS BytesMessage
public void handleMessage(BytesMessage message) {...}
```

```
//receive a JMS MapMessage
public void handleMessage(MapMessage message) {...}

//receive a JMS ObjectMessage
public void handleMessage(ObjectMessage message) {...}

//receive a JMS StreamMessage
public void handleMessage(StreamMessage message) {...}
```

Using this technique gives you access to the JMS `Message` object, allowing you to interrogate the message and extract the header and application properties from the message:

```
public class SimpleJMSReceiver {

   public void handleMessage(TextMessage message) {
      String text = message.getText();
      String username = message.getStringProperty("username");
      String msgId = message.getJMSMessageID();

      //process text message
   }

}
```

Custom message handler method

Of course, you do not need to restrict your POJO message listeners to only the default `handleMessage` methods. In fact, any method in your POJO message listener can be the listener handler method, providing the method contains a single parameter containing either a JMS `Message` type or one of the four message conversion objects (`String`, `byte[]`, `Map`, or `Object`). To use your own method as a message handler, you must set the `defaultListenerMethod` property on the `MessageListenerAdapter` to the name of the method you are using as the message handler. You must also specify whether the `MessageListenerAdapter` should convert the message body or deliver the standard JMS `Message` type to the message handler method. For example, to configure the `MessageListenerAdapter` to use a custom `createTradeOrder` method in the `TradeOrderManager` class that receives a `String` object containing an XML trade order, you would set the `defaultListenerMethod` property to `createTradeOrder` and use the `SimpleMessageConverter`:

```
<bean id="messageListener"
      class="org.springframework.jms.listener.adapter.MessageListenerAdapter">
   <constructor-arg>
      <bean class="TradeOrderManager"/>
   </constructor-arg>
   <property name="defaultListenerMethod" value="createTradeOrder"/>
</bean>
```

In the POJO message listener, you would code the `createTradeOrder` method to receive a `String` object message as follows:

```
public class TradeOrderManager {

   public void createTradeOrder(String xml) {

      //process trade order xml message
      ...
   }
   ...
}
```

If you study the code just shown, you will see that this POJO has no awareness of messaging at all. This is the true value of message-driven POJOs—the ability to abstract the messaging and communication logic away from the POJO, allowing that POJO to focus on business logic rather than messaging infrastructure logic. The above class can now be used both inside and outside a messaging context, allowing it to be tested (or used) outside of the messaging framework.

Message Conversion Limitations

While there are many clear advantages to using automatic message conversion, it is important to understand the limitations of using it as well. Using message conversion increases the overall testability of your receiver class and decouples the POJO message listener from JMS. For example, the following class is configured and used as a message-driven POJO, but can be also used (and tested!) outside of the context of messaging:

```
public class TradeOrderManager {

   public void createTradeOrder(String xml) {

      //process trade order xml message
      ...
   }
   ...
}
```

While this is certainly an attractive feature of Spring's messaging framework, it does have its limitations. For example, when using message conversion, only the body of the message is delivered to the message handler method, meaning that you do not have access to the message headers or application properties. Thus, you cannot extract and use application properties or message header properties in your handler method.

Alternatively, using the JMS `Message` approach in the MDP allows you to gain access to the entire message:

```
public class TradeOrderManager {

   public void createTradeOrder(TextMessage msg) {
      String xml = msg.getText();
      int priority = msg.getJMSPriority();

      //process trade order xml message
      ...
```

```
    }
    ...
}
```

Unfortunately, in this code the TradeOrderManager is tightly bound to JMS and difficult to test outside of the context of JMS. It is important to understand these differences when making the design choice between using message conversion or JMS Message object processing.

The Spring JMS Namespace

Version 2.5 of the Spring Framework introduced the JMS XML namespace support, which greatly simplifies the configuration of message-driven POJOs. To add JMS namespace support to your configuration, you would simply specify the JMS schema in the <beans> element of your Spring application context XML file:

```
<?xml version="1.0" encoding="UTF-8"?>
<beans xmlns="http://www.springframework.org/schema/beans"
       xmlns:xsi="http://www.w3.org/2001/XMLSchema-instance"
       xmlns:jms="http://www.springframework.org/schema/jms"
       xsi:schemaLocation="
http://www.springframework.org/schema/beans
http://www.springframework.org/schema/beans/spring-beans-2.5.xsd
http://www.springframework.org/schema/jms
http://www.springframework.org/schema/jms/spring-jms-2.5.xsd">
```

Without the JMS namespace you must define a separate DefaultMessageListenerContainer bean for each message listener you define. In the following example, two message listeners are defined (messageListener1 and messageListener2), which listen on queue1 and queue2, respectively. Notice how you need to define two message listener containers, one for each message listener. This can get quite cumbersome and verbose when using multiple message listeners:

```
<bean id="messageListener1"
      class="org.springframework.jms.listener.adapter.MessageListenerAdapter">
   <constructor-arg>
      <bean class="SimpleJMSReceiver1"/>
   </constructor-arg>
   <property name="defaultListenerMethod" value="processRequest"/>
</bean>

<bean id="messageListener2"
      class="org.springframework.jms.listener.adapter.MessageListenerAdapter">
   <constructor-arg>
      <bean class="SimpleJMSReceiver2"/>
   </constructor-arg>
   <property name="defaultListenerMethod" value="processRequest"/>
</bean>

<bean id="jmsContainer1"
      class="org.springframework.jms.listener.DefaultMessageListenerContainer">
  <property name="connectionFactory" ref="queueConnectionFactory"/>
```

```
    <property name="destinationResolver" ref="destinationResolver"/>
    <property name="concurrentConsumers" value="3" />
    <property name="destinationName" value="queue1"/>
    <property name="messageListener" ref="messageListener1" />
</bean>

<bean id="jmsContainer2"
      class="org.springframework.jms.listener.DefaultMessageListenerContainer">
    <property name="connectionFactory" ref="queueConnectionFactory"/>
    <property name="destinationResolver" ref="destinationResolver"/>
    <property name="concurrentConsumers" value="3" />
    <property name="destinationName" value="queue2"/>
    <property name="messageListener" ref="messageListener2" />
</bean>
```

With the JMS namespace support, you can combine multiple message listeners within the same message listener container:

```
<bean id="messageListener1" class="SimpleJMSReceiver1"/>
<bean id="messageListener2" class="SimpleJMSReceiver2"/>

<jms:listener-container connection-factory="queueConnectionFactory"
                        destination-resolver="destinationResolver"
                        concurrency="3">
   <jms:listener destination="queue1" ref="messageListener1" />
   <jms:listener destination="queue2" ref="messageListener2" />
</jms:listener-container>
```

This technique is a significant improvement over the older approach.

<jms:listener-container> Element Properties

There are 12 attributes that can be set on the `<jms:listener-container>` element. The most popular attributes you will likely need to use for most mainstream JMS applications are the `connection-factory`, `destination-resolver`, `message-converter`, and `concurrency`. These, as well as the other attributes, are defined here:

`container-type`
: This optional attribute contains the type of message listener container. The possible values are `default` and `simple`, which correspond to the `DefaultMessageListenerContainer` and the `SimpleMessageListenerContainer`. The default value is `default`, indicating that the `DefaultMessageListenerContainer` will be used.

`connection-factory`
: This optional attribute contains a reference to the queue connection factory or topic connection factory bean, and corresponds to the `connectionFactory` property on the `DefaultMessageListenerContainer`.

`destination-resolver`
: This optional attribute contains a reference to the JNDI destination resolver used to resolve JNDI destination names. It points to the bean that is using the

JndiDestinationResolver class, and corresponds to the destinationResolver property on the DefaultMessageListenerContainer.

`message-converter`
: This optional attribute contains a reference to a message converter that is used to convert a JMS Message object into a corresponding Java object. If not specified, the listener container will use the default message converter (SimpleMessageConverter). If you do not want the message listener container to use a message converter, then this attribute must be specified as null. This attribute corresponds to the messageConverter property on the MessageListenerAdapter.

`concurrency`
: This optional attribute corresponds to the concurrentConsumers property on the DefaultMessageListenerContainer and specifies the number of concurrent listener threads the message listener container should start for each listener. If the order of messages in the queue must be preserved, then this attribute should be set to 1 (which is the default value). Note that this value applies to all listeners specified under the message listener container.

`cache`
: This optional attribute identifies the level of caching that the message listener container should use for JMS resources. The possible values are none, connection, session, consumer, or auto. If an external transaction manager (i.e., JTA Transaction Manager) is specified, then the default value is none. However, if no external transaction manager is specified, then the default value is auto, which in most cases defaults to consumer.

`client-id`
: This optional attribute is used for durable subscribers and represents the JMS client ID for the message listener container.

`destination-type`
: This optional attribute indicates the type of JMS destination that is used by this message listener container. The possible values are queue, topic (nondurable subscription), and durableTopic (durable subscription). Since the default value is queue, this attribute needs to be specified only when using the publish-and-subscribe model (pub/sub).

`acknowledge`
: This optional attribute contains the JMS acknowledgment mode. The possible values are auto, client, dups-ok, and transacted. These four values correspond to the AUTO_ACKNOWLEDGE, CLIENT_ACKNOWLEDGE, DUPS_OK_ACKNOWLEDGE, and SESSION_TRANSACTED settings found in the javax.jms.Session interface. The default value is auto.

`transaction-manager`
: This optional attribute contains a reference to an external transaction manager defined through the PlatformTransactionManager bean. If this attribute is specified, then the acknowledgment mode is automatically set to transacted.

`prefetch`
: This optional attribute is used to specify the number of messages to load in a single session.

`task-executor`
: This optional attribute is used to specify the `TaskExecutor`, which is used to run and manage the listener threads. If not specified, the default task executor is used, which is the `SimpleAsyncTaskExecutor`. There are several other task executors available in Spring, including the `SyncTaskExecutor`, `TimerTaskExecutor`, `ThreadPoolTaskExecutor`, and the `WorkmanagerTaskExecutor`. Please refer to the Spring API documentation for a description of each of these task executors.

<jms:listener> Element Properties

Out of the seven attributes that you can specify for the `<jms:listener>` element, only two are required: `destination` and `ref`. The attributes used with this element are defined here:

`destination`
: This required attribute is used to specify the JMS destination name, which is resolved through the destination resolver defined in the message listener container. With the JMS namespace support, you can define multiple listeners using different queues within the same message listener container.

`ref`
: This required attribute contains a reference to the bean that is used as the message listener class.

`method`
: This optional attribute contains the name of the method that is used as the message handler. This attribute can be ignored for message listeners that implement the `MessageListener` or `SessionAwareMessageListener` interfaces.

`selector`
: This optional attribute contains a `String` value representing a message selector that should be applied to this message listener.

`subscription`
: This optional attribute contains the name of a durable subscriber. It is only needed when using the publish-and-subscribe messaging model where the message listener is a durable subscriber.

`id`
: This optional attribute contains the name of the message container bean that this listener is defined under. If you do not specify a bean name, one will be automatically generated.

`response-destination`
: This optional attribute contains a default `ReplyTo` queue in the event that the `JMSReplyTo` header is not set or is not accessible in the receiver. It is typically used when a listener method returns a value other than void.

CHAPTER 10

Deployment Considerations

An enterprise application's performance, scalability, and reliability should be among the foremost concerns in a real deployment environment. The underlying messaging middleware is critical to that environment.

Performance, Scalability, and Reliability

Performance and scalability are terms commonly used together, but they are not interchangeable. *Performance* refers to the speed at which the JMS provider can process a message through the system from the producer to the consumer. *Scalability* refers to the number of concurrently connected clients that a JMS provider can support. When used together, the terms refer to the effective rate at which a JMS provider can concurrently process a large volume of messages on behalf of a large number of simultaneously connected producers and consumers. The distinction between performance and scalability, as well as the implications of what it means to combine them, is very important. A simple test using one or two clients will differ drastically from a test using hundreds or thousands of clients. The following section is intended to be used as a guide to help with performance and scalability testing.

Determining Message Throughput Requirements

Before you embark on your performance and scalability testing effort, consider what you are trying to accomplish. Since any particular vendor may do well with one scenario and not so well in others, the makeup of your application is important to define. Here are some key things to consider:

- The potential size of the user community for your application. While this may be hard to project, it is important to try to predict how it will grow over time.
- The average load required by the application. Given a total size of the user community for your application, how many are going to be actively using it at any given time?

- The peak load required by the application. Are there certain times of the day, or certain days in a month, when the number of concurrent users will surge?
- The number of JMS client connections used by the application. In some cases, the number of JMS clients does not correspond to the number of application users. Middleware products, such as EJB servers, share JMS connections across application clients, requiring far fewer JMS client connections than other applications. On the other hand, some applications use multiple JMS connections per client application, requiring more JMS client connections than users. Knowing the ratio of users to JMS clients helps you determine the number of messages being processed per client.
- The amount of data to be processed through the messaging system over a given period of time. This can be measured in messages per second, bytes per second, messages per month, etc.
- The typical size of the messages being processed. Performance data will vary depending on the message size.
- Any atypical message sizes being produced. If 90 percent of the messages being processed through the system are 100 bytes in size, and the other 10 percent are 10 megabytes, it would be important to know how well the system can handle either scenario.
- The messaging model to be used and how it will be used. Does the entire application use one p2p queue? Are there many queues? Is it pub/sub with 1,000 topics? One-to-many, many-to-one, or many-to-many?
- The message delivery modes to be used. Persistent? Nonpersistent? Durable subscribers? Transacted messages? A mixture? What is the mixture?

Testing the Real-World Scenario

Any vendor can make any software product run faster, provided the company has the right amount of time, proper staffing, commitment, and enough hardware to analyze and test a real-world deployment environment.

The simplest scenario for a vendor to optimize is the fastest performance throughput possible with one or two clients connected. This is also the easiest scenario to test, but it is not the kind of testing we recommend; for one thing, it's difficult to imagine a realistic application that only has one or two clients. More complex testing scenarios that better match your system's real-world environment are preferable.

It is important to know ahead of time if the vendor you have chosen will support the requirements of your application when it is deployed. Because JMS is a standard, you may switch JMS vendors at any time. However, you may soon find yourself building vendor-specific extensions and configurations into your application. It's always possible to change vendors, if you're willing to expend some effort. However, if you wait to find out whether or not your application scales, you may no longer be able to afford the time to switch to another vendor.

This is not intended to imply that using JMS is a risky proposition. These same issues apply to any distributed infrastructure, whether third-party or home-grown, whether it is based on a MOM or based on CORBA, DCOM, EJB, or RMI, and whether it is based on an established vendor or an emerging one. Everything should be sized and tested prior to deployment.

Testing with one client

The most important thing to realize is this:

```
performanceWithOneClient != performanceWithManyClients;
```

Many issues come into play once a message server starts to scale up to a large number of clients. New bottlenecks appear under heavy load that would never have occurred otherwise. Examples include thread contention, overuse of object allocation, garbage collection, and overflow of shared internal buffers and queues.

A vendor may have chosen to optimize message throughput with hundreds or thousands of concurrent clients at the expense of optimizing for throughput with one client. Likewise, a vendor may have optimized for a small number of clients at the expense of scalability with larger client populations.

The best approach is to start with something small and build up the number of clients and the number of messages in increments. For example, run a test with 10 senders and 10 receivers, and 100,000 messages. Next try 100 senders and 100 receivers, and run a test with 1,000,000 messages. Try as many clients as you can, within the limitations of the hardware you have available, and watch for trends.

Send rate versus receive rate

It is extremely important to measure both the send rates and the receive rates of the messages being pumped through the messaging system. If the send rate far exceeds the receive rate, what is happening to the messages? They are being buffered at the JMS provider level. That's OK, right? That is what a messaging product does—it queues things. In some cases, that may be acceptable based on the predetermined throughput requirements of your application and the predictable size and duration of the surges and spikes in the usage of the application. If these factors are not extremely predictable, it is important to measure the long-term effects of unbalanced send and receive rates.

In reality, everything has a limit. If the send rate far exceeds the receive rate, the messages are filling up in-memory queues and eventually overflowing the memory limits of the system, or perhaps the in-memory queues are overflowing to disk storage, which also has a limit. The closer the system gets to its hardware limits, the more the JVM and the operating system thrash to try to compensate, further limiting the JMS provider's ability to deliver messages to its consumers.

Determining hardware requirements

The hardware required to perform testing varies from vendor to vendor. You should have the hardware necessary to do a full-scale test or be prepared to purchase the hardware as soon as possible. If the JMS provider's deployment architecture uses one or more server processes (as in a hub and spoke model), then a powerful server (like a quad-processor) and lightweight clients are appropriate. If the vendor's architecture requires that the persistence and routing functionality be located on the client machine, then many workstations may be required.

If you have limited hardware for testing, do the best you can to run a multiclient load test within the limitations of your hardware. You typically won't see any reasonably indicative results until you have at least 20 JMS clients. You must therefore be able to find a machine or a set of machines that can handle at least that much.

Assuming your client population will be large, truly indicative results start showing up with over 100 JMS clients. Your goal should be to use as many clients as possible within the limits of the testing hardware and to see whether the message throughput gets better or gets worse. A good guideline is to stop adding clients when the average resource utilization on each test machine (both clients and servers) approaches 80 percent CPU or memory use. At 80 percent, you realistically measure the throughput capacity of the JMS provider for a given class of machine and eliminate the possibility of having exceeded the limits of your hardware.

If the CPU or memory utilization does not approach its maximum, and the message throughput does not continue to improve as you add clients, then the bottleneck is probably disk I/O or network throughput. Disk I/O is most likely to be the bottleneck if you are using persistent messaging.

Finding or building a test bed

Building a test bed suitable for simulating a proper deployment environment itself can be a moderately sized effort. Most JMS vendors provide a performance test environment freely downloadable from their web site. In most cases, they provide a test bed sufficient for testing with one client. Alternatively, you can build your own custom test bed or use an open source or commercial performance testing tool.

Long duration reliability

Testing your application over a long period of time is very important. After all, it is expected to perform continuously once deployed. Verifying that the middleware behaves reliably is the first step toward ensuring long-term application reliability.

Once you have a multiclient test bed in place, try running it for an extended period of time to ensure that the performance throughput is consistent. Start out by running the test bed while you can be there to monitor the behavior. Any long-term trends are likely to be detected in the first few hours. Things to watch for are drops in performance throughput, increase in memory usage, CPU usage, and disk usage. When you feel comfortable with the results you see, you may progressively start running tests overnight, over the weekend, or over a week.

Memory leaks

The term "memory leak" refers to a condition that can happen when new memory gets allocated and never freed over a period of time, usually through a repeated operation, such as repeatedly pumping messages through a messaging system. Eventually, the system runs out of available memory; as a result, it will perform badly and may eventually crash.

Although Java has built-in memory management through garbage collection, it is an oversimplification to think that garbage collection permanently solves the problem of memory links. Garbage collection works effectively only when the developer follows explicit rules for the scoping of Java objects. A Java object can be garbage collected only if it has gone out of scope and there are no other objects currently referencing it. Even the best code can contain memory leaks if the developer has mistakenly overlooked an object reference in a collection class that never goes out of scope.

Therefore, you need to monitor for memory leaks during testing. Memory that leaks in small increments may be not be noticeable at first, but eventually these leaks could seriously impact performance. To detect them quickly, it helps to use a memory leak detection tool like OptimizeIt! or JProbe. Even if the JMS provider and other third-party Java products you are using contain obfuscated classes, tools like these still help you prove that your memory requirements are growing (possibly the result of a memory leak), which is a good start.

To Multicast or Not to Multicast

An increasing number of vendors are releasing products based on IP multicasting. To understand the tradeoffs involved in these products, you need a basic understanding of how the TCP/IP protocol family works, and how multicasting fits into the bigger picture.* We won't discuss any particular JMS implementations or suggest that one vendor might be better than another; our goal is to give you the tools that you need to ask intelligent questions, evaluate different products, and map out a deployment strategy.

* A comprehensive discussion of TCP/IP networking is out of the scope of this book. If you want detailed treatment of these protocols, see *Internet Core Protocols (http://oreilly.com/catalog/9781565925724/)*, by Eric Hall (O'Reilly). If you're interested in network programming in Java, see *Java Network Programming (http://oreilly.com/catalog/9780596007218/)*, by Elliotte Rusty Harold (O'Reilly).

TCP/IP

TCP/IP is the name for a family of protocols that includes TCP (Transmission Control Protocol), UDP (User Datagram Protocol), and IP (Internet Protocol). The protocols are layered: IP provides low-level services; both TCP and UDP sit "on top of" IP.

TCP is a reliable, connection-oriented protocol. A process wishing to establish communication with one or more processes across a network creates a connection to each of the other processes and sends and receives data using those connections. The network software, rather than the application, is responsible for making sure that all the data arrives, and that it arrives in the correct order. It takes care of acknowledging that data has been received, automatically discards duplicate data, and performs many other services for the application. If something happens with the connection, the process on either side of the connection will know *almost* immediately that the connection has been permanently broken.†

Most high-level network protocols (and most JMS implementations) are built on top of TCP, for obvious reasons: it's a lot easier to use a protocol that takes care of reliability for you. However, reliability comes with a cost: a lot of work is involved in setting up and tearing down connections, and additional overhead is required to acknowledge data that's sent and received. Therefore, TCP is slower than its unreliable relative, UDP.

UDP

UDP (User Datagram Protocol) is an unreliable protocol; you send data to a destination, but there's no guarantee that the data will arrive. If it doesn't arrive, you'll never find out; furthermore, the process receiving the data will never know that you sent anything.

This sounds like a bad basis for reliable software, but it really only means that applications using UDP have to take reliability into their own hands: they need to come up with their own mechanism for verifying that data was received, and for retransmitting data that went astray. In practice, applications that need reliability guarantees can either use TCP, or can incorporate software to build reliability on top of UDP. Most applications have taken the easier route, but a few important applications (like DNS and the early versions of NFS) make extensive use of UDP.

IP Multicast

The simplicity of UDP makes possible a kind of service that's completely different from anything in the TCP world. Because it is connection-oriented, TCP is fundamentally limited to point-to-point communications. UDP offers the notion of a "multicast," in which an application can send data to a group of recipients. Multicasting is based on

† If a connection is not sending or receiving any data, it could take a while before the owning process is signaled about a problem, depending on the network settings.

a special class of addresses, known as Class D addresses.‡ Class D addresses are not assigned to individual hosts; they're assigned to multicast groups. Hosts can join and leave groups that they have an interest in. Data sent to a multicast address will only be received by the hosts in the multicast group. At least from the network's standpoint, multicast is much more efficient when you need to send a message to many recipients.

Multicasting maps naturally into the sorts of things we want messaging systems to do. Many messaging products use multicasting for one-to-many pub/sub broadcast of messages. Most have built some level of reliability on top of UDP. If this issue is important to you, it would be in your interest to delve deeper and find out exactly what your JMS vendor has, or has not, implemented. Multicast has its drawbacks as well. UDP traffic is usually not allowed through a firewall, so you may have to negotiate with your network administrators or find some workaround if you need to get multicast traffic through your company's firewalls. Furthermore, multicast relies heavily on special routing software. Most modern routers support multicast, but lots of old routers are still in service. Even if you have up-to-date routers within your corporate network, and your network administrators know how to configure multicast routing, there's still the Internet; multicasting does not realistically work across the Internet (this is discussed in more detail later in the chapter). As a configuration and maintenance consideration, multicast addresses must be coordinated across the network to avoid collisions. These drawbacks are especially important if you are building an application that you want to sell to others, who in turn expect to deploy it easily.

Messaging Over IP Multicast

In the following section we will explore the tradeoffs of using messaging over an IP multicast architecture. It is important for you to understand the issues as you map out your deployment strategy.

Duplication, ordering, and reliability of messages

If a messaging vendor wishes to provide full reliability for IP multicast and UDP, it must build TCP-like semantics into the JMS provider layer to compensate for duplicate datagrams, out of order datagrams, and datagrams that could never possibly get to the intended destination. Either the JMS provider has to incur the overhead of detecting and compensating for duplicate datagrams or the application needs to be tolerant of duplicate messages. If the duplication of datagrams is not dealt with at the JMS provider level, it is only really viable for `DUPS_OK_ACKNOWLEDGE`. No matter what, a messaging vendor has to implement the reliability necessary to ensure guaranteed ordering, since UDP doesn't ensure that packets are received in the same order that they are sent.

‡ A Class D network address is one defined as having the range of 224.0.0.0 through 239.255.255.255. Class D network addresses are reserved for IP multicast.

A messaging vendor should support some sort of error detection to know when a UDP datagram is lost. Ideally it should know that a client can't be reached due to a network boundary across an unsupported network router. The JMS specification allows for a nondurable JMS subscriber to miss messages, but is intentionally vague about this since it is not a goal of the specification to impose an architecture on a JMS provider. However, for all practical purposes, nonguaranteed messaging means that messages *may* be lost, and that should mean they may only be lost once in a while. For both cases, some sort of acknowledgment semantics are required.

Centralized and decentralized architectures

A TCP-based messaging system generally uses a hub-and-spoke architecture whereby a centralized message server, or cluster of message servers, communicates with JMS clients using TCP/IP, SSL, or HTTP connections. The centralized server is responsible for knowing who is publishing and who is subscribing at any given time. Message servers may operate in a cluster spread across multiple machines, but to the clients there only appears to be a single logical server. Message servers operating in a cluster can intelligently route messages to other servers. Clustering may provide load balancing and may help to optimize network traffic by selectively filtering and routing only the messages that need to get to a particular node. The servers are also responsible for persistence of guaranteed messages, and for Access Control Lists (ACLs) that grant permissions to subscribers on a per-topic basis. The messages are only delivered to the subscribers that are interested in a particular topic, and only to those that have the permissions to get them. A centralized server also makes it easier to add subscribers: when a new subscriber comes online, only the message server needs to know about it.

At the same time, a centralized architecture may introduce a single point of failure: if the main server in a cluster (the server to which clients initially connect) goes down, the entire cluster may become unavailable. A JMS provider may solve this problem by distributing the connections across multiple servers in the cluster. If one server goes down, the other servers can continue to operate, thus minimizing the impact of the failure. Reconnect logic may also be built into the client, enabling it to find another server if its initial server goes down.

Multicasting implies a drastically different architecture, in which there usually is no centralized server. Because there is no central server, there is no single point of failure; each JMS client broadcasts directly to all other JMS clients. One consequence of this architecture is that every publisher and every subscriber may have local configuration information about every other JMS client on the system. This can be an extremely important consideration for deployment administration. In the absence of a higher-level administrative framework, local configurations have to be updated on every client whenever a new client or a new topic is added.

A decentralized architecture may also mean that the persistence mechanism for guaranteed messaging is pushed out to the client machines. No matter how efficient the storage algorithm, disk I/O is always going to be the biggest bottleneck. Choosing to use such an architecture would require that the client machines have disk storage that is both fast and large.

There is disagreement as to whether guaranteed messaging (storing persistent messages) benefits from a decentralized architecture. Proponents of a decentralized architecture argue that the I/O load is distributed among the clients and is therefore faster. On the other hand, client I/O is not nearly as reliable, nor is it as fast as a centralized server with a powerful disk system.

Network routers and firewalls

Although technically possible, it is unlikely that a firewall administrator will allow UDP traffic to pass through a firewall. Firewalls typically disallow all traffic, except for traffic to or from specific hosts, using specific protocols. UDP traffic is rarely allowed through a firewall for various reasons.

In recognition of the problems with IP multicast (lack of support and firewall blocking), messaging vendors that use IP multicast provide software bridge processes to carry messaging traffic across routers and firewalls. The bridges may consist of one or more processes connected together by HTTP, SSL, or TCP/IP.

If you're considering a vendor that supports multicasting, it is worth considering what percentage of your message traffic is going through one of these bridges. If all of your messages are going through the firewall over an SSL or HTTP connection, there will be little point in using multicasting behind the firewall for performance reasons. If the routers in your deployment environment require that a number of TCP/IP-based bridges be put in place, the performance benefits of multicast are diminished, depending on how many of these you have to put in place and administer. The messaging system is only as fast as its slowest link.

If most of the message traffic is confined to your corporate LAN or a VPN and you have full control over it, IP multicasting is a very attractive option.

Some vendors support both centralized and decentralized architectures

In recognition of these issues, the vendors who support IP multicast also provide centralized servers using TCP/IP socket connections. This could mean you have two different architectures to configure and support: one configuration for the nonguaranteed one-to-many pub/sub multicast of messages within a subnet on your corporate LAN, and another for everything else. It is important to consider what it will mean to choose one of these architectures at deployment time or how you will switch from one mode to the other after your application is deployed.

The Bottom Line

IP multicast has significant network throughput benefits in a one-to-many broadcast of information. A single multicast message to multiple recipients will always cause less network traffic than sending the message to each recipient via a TCP connection. A messaging vendor picks and chooses how much reliability to build on top of UDP based on the quality of service required for the message as defined by JMS.

However, the choice is not that simple when it is applied to a deployment environment in a messaging product. The performance advantages of IP multicasting are only viable for a certain deployment environment. These advantages can diminish depending on the types of messages in your application, the networking hardware at your site, the deployment environment (intranet, extranet, Internet), and the complexity of administration.

Make sure to benchmark your application carefully before making a final decision, using the guidelines we discussed earlier in this chapter. You may be surprised at what you see. When a JMS provider is put under heavy stress with lots of clients, there are so many other factors involved that the speed at which network packets go across the wire is not usually a significant factor. You may see that one vendor's implementation of messaging over IP multicast will perform vastly differently from another's—even with the use of nonguaranteed messaging. You may even find that one vendor's TCP-based implementation performs better than another vendor's multicast implementation.

Security

In this section, we are only going to concern ourselves with those aspects of security that are commonly supported by JMS providers. You need to think about three aspects of security: authentication, authorization, and secure communication. How these aspects of security are implemented is vendor-specific and each vendor uses its own combination of available technologies to authenticate, authorize, and secure communication between JMS clients.

We will also discuss firewalls and HTTP tunneling as a solution to restrictions placed on JMS applications by organizations.

Authentication

Simply put, authentication verifies the identity of the user to the messaging system; it may also verify the identity of the server to the JMS client. The most common kind of authentication is a login screen that requires a username and a password. This is supported explicitly in the JMS API when a `Connection` is created, as well as in the JNDI API when an `InitialContext` is created. JMS providers that use username/password authentication may support either of these solutions:

```
Properties env = new Properties();

env.put(Context.SECURITY_PRINCIPAL, "username");
env.put(Context.SECURITY_CREDENTIALS, "password");
```

```
InitalContect ctx = new InitialContext(env);

TopicConnectionFactory factory =
    (TopicConnectionFactory)ctx.lookup("...");

TopicConnection connection =
    factory.createTopicConnection("username", "password");
```

JMS providers may also use more sophisticated mechanisms for authentication, such as secret or public key authentication. Secret key authentication, most commonly used in Kerberos, requires the participation of a Kerberos server.§ Public key authentication, most commonly used in SSL, is based on a chain of certifying authorities. Each of these systems has its supporters and detractors, but the end result is the same: the connecting client is given permission to access the system.

Authorization

Authentication is only the first step in the security process, but it's the basis for what follows. Once you have verified the identify of the user, you can make intelligent decisions about what that user is allowed to do. That's where authorization comes in. Authorization (a.k.a. access control) applies security policies that regulate what a user can and cannot do within a system. Authorization policies are usually set up as access control lists by the system administrators. Authorized users are given an identity in the system and assigned to user groups, which may themselves be a part of a larger group. Groups and individual users (identities) are assigned permissions dictating which topics, queues, or connection factories they are allowed to access. Permissions may be configured to grant all members of a group access except for some specified members, or deny all members of a group except some specified members. Some JMS providers may choose to check access control lists on every message delivered, while others simply control the destinations or connection factory that a JMS client can obtain from the JNDI namespace. Generally, authorization policies work better in a centralized messaging system, since it can be centrally managed.

Most JMS providers provide hierarchical topic trees that allow consumers to subscribe to different levels of topics using wildcard substitution. For example, topics could be divided into "ACME.SALES.SOUTHWEST.ANVILS" and "ACME.SALES.NORTHEAST.ANVILS." A subscriber can subscribe to "ACME.SALES." and see all the messages published for all the sales of ACME, though that may not be the desire of the system administrator. A companion security feature allows permissions to be set at each level in the topic tree, thus making access control much easier to manage by providing more finely grained access control.

§ Although a system may use Kerberos to authenticate a user, the system will probably use SSL for secure communications.

Secure Communication

Communication channels between a client and a server are frequently the focus of security concerns. A channel of communication can be secured by physical isolation (like a dedicated network connection) or by encrypting the communication between the client and the server. Physically securing communication is expensive, limiting, and pretty much impossible on the Internet, so we will focus on encryption. When communication is secured by encryption, the messages passed are encoded so that they cannot be read or manipulated while in transit. This normally involves the exchange of cryptographic keys between the client and the server. The keys allow the receiver of the message to decode and read the message.

There are two basic ways that messages are encrypted by JMS providers today: SSL and Payload Encryption. SSL (Secure Socket Layer) is an industry-standard specification for secure communication used extensively in Internet applications. With SSL, the JMS provider's protocol is encrypted, protecting every aspect of the JMS client's exchanges with the message service. Payload Encryption allows messages to be encrypted on a per-topic, per-queue basis. This unusual variance minimizes overhead by encrypting only the messages that need it, rather than everything on the whole connection.

For example, a `PricingServer` class may not need to encrypt the broadcast of updated stock prices since that same information is being replicated to every subscriber with an authenticated connection. The response message with a "Buy Trade Order" would more likely be encrypted since that is sensitive data that is unique to each subscriber. Payload Encryption can also ensure end-to-end security between a producer and a consumer. Without it, there may be nothing preventing a sender from connecting to the message server using a SSL connection and receiving an unencrypted message using a non-SSL connection.

Firewalls and HTTP Tunneling

Firewalls are systems that serve as the gateway between an organization and a broader network such as the Internet. These gateways filter all incoming and outgoing messages. Firewalls only allow packets of a predetermined type and protocol to pass between computers within the organization and those in the broader network. Firewalls help to stop malicious attacks against an organization's information systems by outside parties.

In most cases, firewalls allow HTTP traffic to flow without restriction. Since HTTP is not the native protocol of most JMS providers, JMS providers must piggy-back their protocol on top of HTTP to penetrate a firewall and exchange messages. This is commonly referred to as *HTTP tunneling*. HTTP tunneling is not really complicated. It involves nesting a JMS provider's native protocol inside HTTP requests and responses. Because the JMS provider's protocol is nested in HTTP, it's hidden from the firewall and effectively tunnels through unnoticed.

In any JMS application that must communicate across a variety of firewalls with large user populations, HTTP tunneling is a necessity. This is especially true when the clients are not centrally managed and may be added and removed at will, which is often the case in B2B applications.

The level of support for tunneling varies, depending on the JMS provider. In addition to tunneling through server-side firewalls, it is important to know if the JMS client can tunnel through a client-side firewall and if HTTP proxies are supported. It is also important to know if the vendor supports HTTP 1.1 Persistent Connections, HTTP 1.0 Keep-Alive Connections, or simple HTTP 1.0 Connections.

Connecting to the Outside World

There are often entities outside your corporation that you need to interact with. You may have trading partners, financial institutions, and vertical business portals to connect to and communicate with. These outside entities usually have established protocols that they already use for electronic communication. An Electronic Data Interchange (EDI) system may have nightly batch jobs that export flat files to an FTP site. A trading partner may expect to send and receive HTTP transmissions as its way of communicating with the outside world. A supply chain portal may require that you install one of their clients on your site in order to communicate with them through whatever protocol they dictate. Sometimes email is required as a way of sending a "Thank you for your order" message.

Ideally each of these outside entities would have a close working relationship with you and would allow you to install a JMS client at each site. That would make communication very easy—but it's not how the world works. These other communication mechanisms may have been in place for a number of years and their users aren't about to rip them out just because you want them to. They may not be capable of changing the way their systems work just to accommodate your JMS provider. These are "legacy systems"; in the future, they may gradually disappear, but for the time being, we have to figure out how to work with them.

There is recent activity in the area of providing a RESTful interface to JMS. REST (Representational State Transfer) is an architecture style outlined in a doctoral dissertation by Roy Fielding that describes a fixed set of verbs (e.g., GET, POST, PUT, and DELETE) and nouns (or system resources) that those verbs act on. As of this writing, ActiveMQ and IBM WebSphere MQ both provide a RESTful interface to JMS, allowing an external client to use a URL to send and receive messages.

One of the issues with developing a RESTful interface to JMS is that it unfortunately does not conform perfectly to the core REST principles. For example, one of the REST principles states that the GET action must be idempotent, meaning that repeated operations against the same action must yield the same results. However, the GET operation on a queue removes a message from the queue, thereby altering the state of the queue and hence returning different results on repeated GET operations.

Another issue with a RESTful interface to JMS is establishing what the definition of a JMS resource is and how it should be used in relation to the REST verbs. For example, what does it mean to DELETE a queue? Does it mean to remove the physical queue or remove a message from the queue? What about a GET on a queue? Does that return a reference to the queue itself or does it return a message from the queue? ActiveMQ and IBM WebSphere MQ both take a slightly different approach to these questions.

While there is promise for this emerging technology, it is still largely inconsistent and unproven at this point. Until then, we are left to building messaging bridges, or connectors, to those other protocols. As illustrated in Figure 10-1, a bridge is simply a JMS client. Its sole purpose is to receive data using the foreign protocol, create a JMS message, and send it along through your JMS-based system. Likewise an outbound connector would listen for messages from your JMS-based system and transmit the message out into the world using the protocol expected by the entity at the other end.

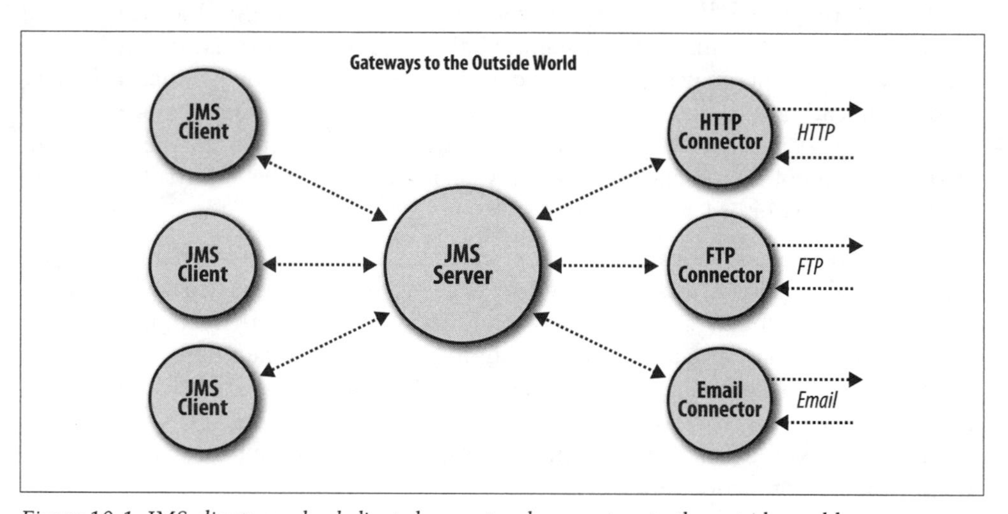

Figure 10-1. JMS clients can be dedicated as protocol connectors to the outside world

The JMS specification does not suggest this notion of connectors.‖ However, legacy systems are a fact of life. In recognition of this, most JMS vendors are starting to provide connectors to legacy systems as a way to provide added value. If your JMS provider does not support the connector you are looking for, it is typically easy enough to write your own. In fact, this is an ideal situation for using `CLIENT_ACKNOWLEDGE` mode. As illustrated in Figure 10-2, a JMS consumer can explicitly acknowledge the receipt of the message once its data transmission has been successfully completed.

‖ The use of the term *connector* in this discussion should not be confused with "connectors" as defined by the J2EE connector specification—a different thing altogether.

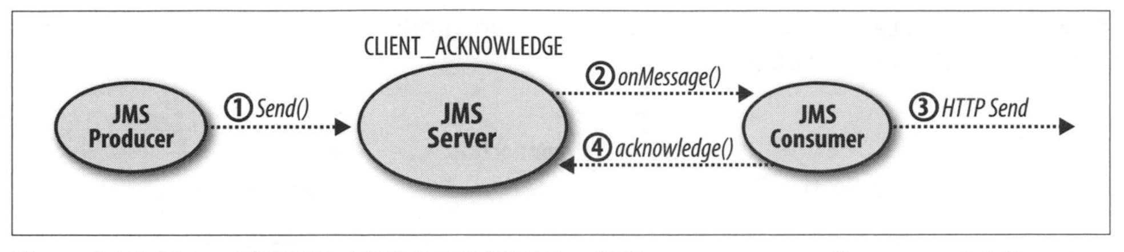

Figure 10-2. Using CLIENT_ACKNOWLEDGE, a JMS consumer can still ensure reliability when bridging to other protocols

It is important to know that end-to-end quality of service may not be guaranteed when using bridges to other protocols. In Figure 10-2, the HTTP send may succeed, yet the `acknowledge()` may fail.

Bridging to Other Messaging Systems

JMS does not provide for interoperability between JMS providers and other non-JMS providers. A JMS client from one vendor cannot talk directly to a JMS server from another vendor. Interoperability between vendors was not a goal of the specification's creators, since the architecture of messaging vendors can be so vastly different. Since messaging is designed to provide an abstraction layer between a message producer and a message consumer, we ought to be able to send a message with a Java client using JMS and receive that message in a receiver written in C++ or C#. This would require either a JMS provider that contained a message bridge or a separate message bridge to convert the JMS message to a message that could be understood by another messaging system (e.g., MSMQ).

IBM's WebSphere MQ and the open source ActiveMQ JMS provider are two such messaging vendors that provide the ability to use both JMS and the vendor's native API, allowing interoperability between languages and platforms. Alternatively, there are open source and commercial messaging bridges available that will perform this function as well. Another solution is to build a connector process that is a client of both providers. Its sole purpose is to act as a pass-through mechanism between the two providers, as shown in Figure 10-3. This is one of the reasons why the message itself is required to be interoperable between vendors. The message need not be recreated as it is passed along to the other JMS client.

It is important when selecting a JMS provider that you consider your interoperability requirements and whether the messaging provider can support those requirements. For example, some JMS providers (including Java EE application servers) can only send and receive JMS messages. These types of JMS providers would not be of much use if your requirements are to provide interoperability between Java and C#. Most vendors that supply a built-in messaging bridge have restrictions in terms of the types of JMS message types that can be sent. For example, using the JMS `ObjectMessage` will, for the most part, restrict your producers and consumers to Java, whereas use of the

`MapMessage` or `BytesMessage` provides maximum portability across most messaging providers and messaging bridges.

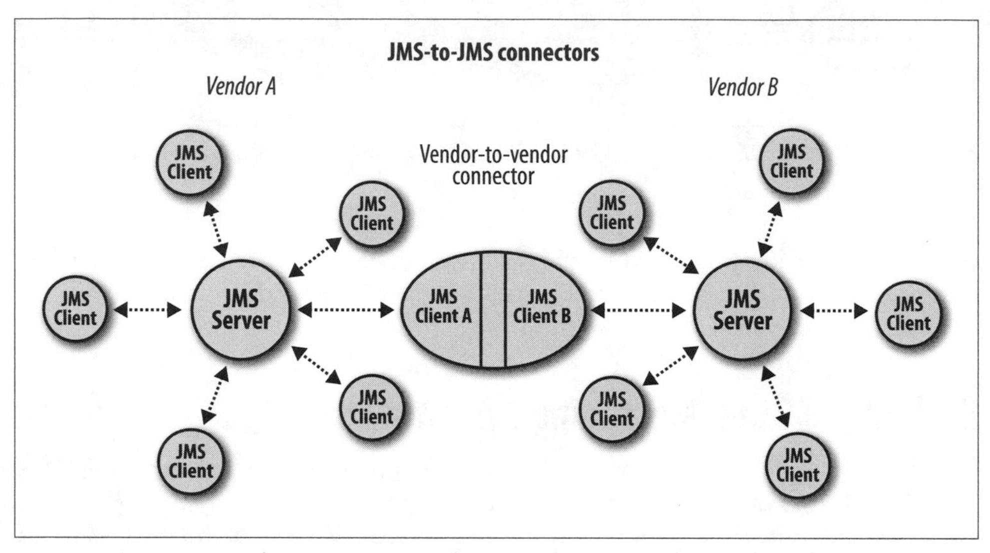

Figure 10-3. Connecting from one JMS provider to another is a simple pass-through process that is a client of both

CHAPTER 11

Messaging Design Considerations

Until now we have been focusing on the concepts and semantics of JMS messaging. This chapter expands on those concepts and introduces some of the design considerations that need to be addressed when building messaging systems.

Internal Versus External Destination

The Java Message Service API consists of a set of standard interfaces that must be implemented by open source or commercial vendor products called *JMS providers*. There are many different types of open source and commercial JMS providers ranging from J2EE application servers (e.g., JBoss, WebLogic, IBM WebSphere, Oracle AS) to standalone solutions (e.g., ActiveMQ, SonicMQ, IBM WebSphere MQ).

Because most Java EE application servers can also serve as JMS providers, a common design choice is deciding whether to leverage your current application server environment or use an external JMS provider. After all, why complicate your architecture and incur additional middleware licensing fees if you can simply utilize your existing application server? While this seems like a straightforward question, there are several implications and limitations to using an application server as a JMS provider. As you will see in this section, using an existing Java EE application server environment versus an external standalone JMS provider is an important design decision that carries with it many implications with respect to an overall messaging solution.

There are two primary deployment topologies with respect to JMS providers: *Internal Destination* and *External Destination*. The Internal Destination topology refers to queues and topics that are administered by an application server (e.g., WebLogic) that also hosts web-based or server-based applications, whereas the External Destination Topology refers to queues and topics that are administered on a dedicated system outside of the context of web-based or server-based applications. The following sections describe the details and design considerations surrounding each of these design choices.

Internal Destination Topology

As stated earlier, the Internal Destination topology refers to queues and topics that are administered by an application server that also hosts web-based or server-based applications. Since all Java EE 4 and above application servers are also JMS providers, this is something that is fairly easy to configure and use. Use of the Internal Destination topology is common for message-driven beans (Chapter 8), but is certainly not a requirement.

As illustrated in Figure 11-1, with the Internal Destination topology queues and topics are administered by the Java EE application server, which also hosts the applications deployed as either WAR (Web Archive) files, EAR (Enterprise Archive) files, or simple JAR (Java Archive) files. External message consumers or receivers must connect to that application server (usually through TCP/IP) to send and receive messages.

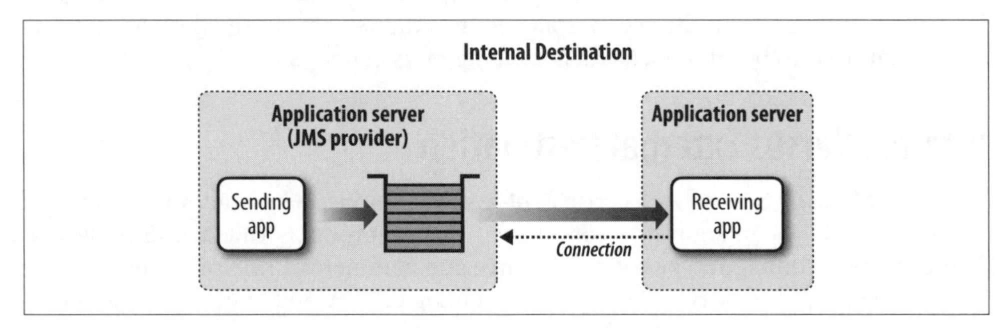

Figure 11-1. Internal Destination

There are a variety of issues associated with the Internal Destination topology that are important to consider. First, using a Java EE application server as the JMS provider restricts message producers and consumers to the Java platform (i.e., JMS) only. This means that your messaging solution will not support heterogeneous messaging clients. While this restriction may not be an issue for you now, it may be an issue with respect to future expansion. With mergers and acquisitions on the rise, many companies find heterogeneous integration an important capability a system must support in order to maintain architectural vitality.

Maintaining a healthy separation of concerns is another issue with the Internal Destination topology. Application servers that play a dual role of being an application hosting server and a messaging provider at the same time can slow down a system and create significant system bottlenecks. A messaging server consumes a significant amount of system resources, particularly with respect to CPU and available thread count. When an application server instance hosts web-based or server-based applications and acts as a JMS provider at the same time, resources needed by the applications may be consumed by message processing, thereby starving applications of much-needed system resources.

Message server availability is another issue associated with the Internal Destination topology. When the application server instance needs to be taken down for system maintenance or application deployment, your messaging system comes down as well, preventing external message producers or consumers from getting to the queues and topics.

Based on these issues, it is generally not a good idea to use an application server as a dual purpose server for both applications and message processing. That said, there are times when it might make sense to use the Internal Destination topology. One of these use cases is when you have a self-contained Java-based application that utilizes messaging to decouple internal components to reduce bottlenecks and increase scalability and throughput. In this scenario, it is unlikely that the queues and topics will be used outside of the context of the application, so using the Internal Destination topology in this case makes sense.

External Destination Topology

With the *External Destination topology*, the JMS provider is deployed as a dedicated server that is separate from any Java EE application servers used to host applications (including messaging producers and consumers). As illustrated in Figure 11-2, the JMS provider server instance is only responsible for managing the queues and topics, leaving the Java EE application servers free to host web-based or server-based applications. This topology supports a healthy separation of concerns between application hosting and providing messaging services, resolving several of the issues encountered with the Internal Destination topology.

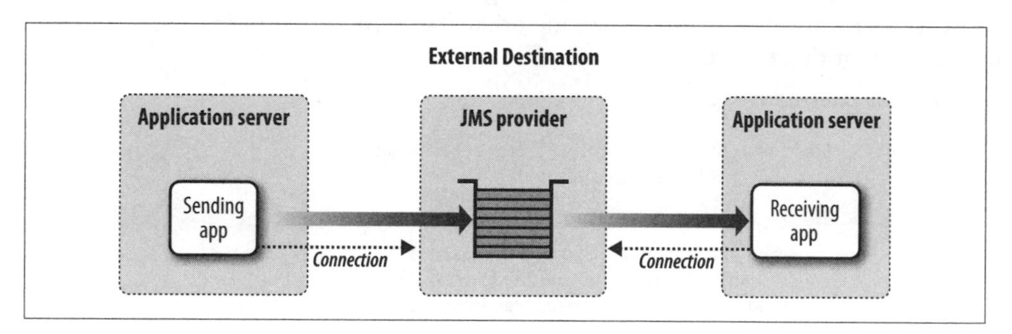

Figure 11-2. External Destination

While the external JMS provider can be deployed using a dedicated Java EE application server instance, the limitation of JMS-only messaging would still apply. For this reason, a standalone JMS provider such as ActiveMQ, SonicMQ, or IBM WebSphere MQ (to name a few) is typically used instead. These messaging providers support the JMS API, but also expose a native API for the programming languages supported by that provider. For example, as of the time of this writing ActiveMQ (a popular open source messaging provider) supports thirteen different languages and platforms, including C, C++, C#, Perl, Ruby, and Smalltalk.

The External Destination topology supports heterogeneous integration through messaging, and also provides a high degree of separation between applications and the JMS provider. For instance, application servers can be taken down for maintenance or application deployment without affecting the rest of the system from a messaging standpoint. Other applications can continue to send and receive messages while application servers are unavailable.

The External Destination topology also allows the JMS provider to be co-located on the same physical machine or deployed on a dedicated physical machine. The choice between where the JMS provider is deployed is largely based on the throughput required by your messaging solution. Systems with low messaging throughput requirements (e.g., 50 messages per second) might be good candidates for co-location, whereas systems with high messaging throughput requirements (e.g., 2,000 messages per second) might warrant a separate physical machine.

Request/Reply Messaging Design

In Chapter 4 we introduced point-to-point messaging using a simple request/reply model. In this scenario the message producer (`QBorrower`) sent a loan request to the message consumer (`QLender`) and waited (blocking wait) for a response from the `QLender` on whether the loan was approved or denied. To implement the request/reply model we used a technique known as *message correlation*, where messages sent to the response queue were correlated with the original message using the `JMSMessageID` and `JMSCorrelationID`. The following are the original `QBorrower` and `QLender` listings used to implement the request/reply scenario.

```
public class QBorrower {
   ...
   public QBorrower(String queuecf, String requestQueue,
                    String responseQueue) {
      try {
         ...
         // Lookup the request and response queues
         requestQ = (Queue)ctx.lookup(requestQueue);
         responseQ = (Queue)ctx.lookup(responseQueue);
      }
      ...
   }

   private void sendLoanRequest(double salary, double loanAmt) {
      try {
         // Create JMS message
         MapMessage msg = qSession.createMapMessage();
         msg.setDouble("Salary", salary);
         msg.setDouble("LoanAmount", loanAmt);
         msg.setJMSReplyTo(responseQ);
```

```
            // Create the sender and send the message
            QueueSender qSender = qSession.createSender(requestQ);
            qSender.send(msg);

            // Wait to see if the loan request was accepted or declined
            String filter =
               "JMSCorrelationID = '" + msg.getJMSMessageID() + "'";
            QueueReceiver qReceiver = qSession.createReceiver(responseQ, filter);
            TextMessage tmsg = (TextMessage)qReceiver.receive(30000);
            if (tmsg == null) {
               System.out.println("QLender not responding");
            } else {
               System.out.println("Loan request was " + tmsg.getText());
            }
         ...
         }
      }
      ...
   }
```

To implement request/reply message processing in the preceding `QBorrower` class we had to create a `Queue` object for the response queue, create a message selector based on the `JMSCorrelationID` header property, and then create a `QueueReceiver` using the response queue and message selector. Then, in the `QLender` class, we had to set the `JMSCorrelationID` header property to the `JMSMessageID` of the original message when sending the reply:

```
   public class QLender implements MessageListener {
      ...
      public void onMessage(Message message) {
         try {
            ...
            // Send the results back to the borrower
            TextMessage tmsg = qSession.createTextMessage();
            tmsg.setText(accepted ? "Accepted!" : "Declined");
            tmsg.setJMSCorrelationID(message.getJMSMessageID());

            // Create the sender and send the message
            QueueSender qSender =
               qSession.createSender((Queue)message.getJMSReplyTo());
            qSender.send(tmsg);
            ...
         }
      }
   }
```

The message correlation code just shown was necessary to ensure that the message being received by the response queue was intended for the loan request originally sent. Keep in mind, other borrower clients may be making loan requests at the same time, creating multiple response messages in the loan response queue.

Another technique for accomplishing the same thing but with far less code is using the `javax.jms.QueueRequestor` class (or `javax.jms.TopicRequestor` class for topic-based request/reply). With the QueueRequestor class, senders and receivers do not have to worry about setting the JMSCorrelation header property or even creating the corresponding QueueReceiver to receive the reply. Instead, the QueueRequestor class creates a unique *temporary queue* whose reference is passed to the message receiver through the JMSReplyTo header property. This temporary queue only has context for the communications between a sender and receiver for a specific request. By using a temporary queue, the QBorrower can be assured that the next available message in that queue is a response to the prior loan request message sent.

Sending a message and waiting (while blocking) for the response is done though a single request method call on the QueueRequestor class. The following revised QBorrower uses the QueueRequestor class to send the loan request and wait for a response:

```
public class QBorrower {

   public QBorrower(String queuecf, String requestQueue) {
      try {
         ...
         // Lookup the request queue
         requestQ = (Queue)ctx.lookup(requestQueue);
         ...
      }
      ...
   }

   private void sendLoanRequest(double salary, double loanAmt) {
      try {
         // Create JMS message
         MapMessage msg = qSession.createMapMessage();
         msg.setDouble("Salary", salary);
         msg.setDouble("LoanAmount", loanAmt);

         QueueRequestor requestor = new QueueRequestor(qSession, requestQ);
         TextMessage tmsg = (TextMessage)requestor.request(msg);
         if (tmsg == null) {
            System.out.println("Lender not responding");
         } else {
            System.out.println("Loan request was " + tmsg.getText());
         }
      }
      ...
   }
   ...
}
```

There are several differences between the modified QBorrower just shown and the original QBorrower from Chapter 4, even though they both do exactly the same thing. First, notice that the QBorrower constructor no longer requires a response queue name argument, nor does it need to do a JNDI lookup on the response queue. Since the QueueRequestor class creates a temporary queue for the loan response message, you no longer have to specify an administered queue to handle responses, nor do you need to manage a separate queue in the JMS provider. The QueueRequestor class will take care of creating and destroying the temporary queue.

Second, notice that the sendLoanRequest method is significantly shorter than the prior version from Chapter 4. As a matter of fact, the act of sending the loan request and waiting for the response has been reduced to only two lines of code:

```
QueueRequestor requestor = new QueueRequestor(qSession, requestQ);
TextMessage tmsg = (TextMessage)requestor.request(msg);
```

The first line constructs a new QueueRequestor object using the QueueSession and loan request queue. The second line then sends the message and automatically blocks and waits for the response. A reference to the temporary queue created by the QueueRequestor class is passed to the receiver (in this case QLender) through the JMSReplyTo message header property.

Now that we are using a temporary queue, we can remove the code in the QLender class that set the JMSCorrelationID. Notice that because the QLender was already agnostic as to the response queue, the other code stays the same:

```
public class QLender implements MessageListener {

   ...
   public void onMessage(Message message) {
      try {
         ...
         // Send the results back to the borrower
         TextMessage tmsg = qSession.createTextMessage();
         tmsg.setText(accepted ? "Accepted!" : "Declined");

         //since we are using a temporary queue, we no longer
         //need to set the JMSCorrelationID property
         //tmsg.setJMSCorrelationID(message.getJMSMessageID());

         // Create the sender and send the message
         QueueSender qSender =
            qSession.createSender((Queue)message.getJMSReplyTo());
         qSender.send(tmsg);
         ...
      }
   }
}
```

While this alternative request/reply technique simplifies the source code and reduces the number of required administered queues, it does have some limitations. First, there is no way to specify a timeout value with the request method on the QueueRequestor as we did with the receive method on the QueueReceiver:

```
//QueueReceiver
TextMessage tmsg = (TextMessage)qReceiver.receive(30000);

//QueueRequestor
TextMessage tmsg = (TextMessage)requestor.request(msg);
```

This is somewhat an issue in that if the message consumer is not responding, the message producer will appear to "hang" and might require a restart. If you did restart the message producer, the original message would still be sitting on the request queue waiting to be processed by the message consumer. Implementing a timeout would require you to override the `javax.jms.QueueRequestor` `receive()` method, thereby possibly creating nonportable JMS code. When considering this approach the benefits of streamlined code should be weighed against the lack of a receive timeout capability.

Another limitation of the `QueueRequestor` is that the `QueueSession` cannot be transacted, and will not support the `CLIENT_ACKNOWLEDGE` message acknowledgment mode. These limitations should also be considered before using the `QueueRequestor` technique.

Messaging Design Anti-Patterns

There are several messaging-related anti-patterns that manifest themselves in production environments. An anti-pattern is a practice that is repeated but produces negative results (unlike a pattern, which is a repeatable process that produces positive results). Three of the most common messaging anti-patterns are the *single-purpose queue*, *message priority overuse*, and *message header misuse*. This section will cover the details of each of these anti-patterns and describe ways to avoid them.

Single-Purpose Queue

A common messaging anti-pattern is designing a system with only a single purpose queue. Typically this problem manifests itself when a single queue handles different types of messages (e.g., book orders, order status requests, and order cancellations), but problems can also occur when a single purpose queue is used for the same type of message (e.g., book orders). We will start with the first scenario since it is most common, and then move onto the second scenario, which is a little more subtle.

Systems that use a single purpose queue often have a single message listener class that acts as a router. The router listener receives the next message on the queue, determines the message type, and redirects processing to some other class to process that message. This design scenario is illustrated in Figure 11-3.

The routing rules used by the listener router can be based on the JMS message type (e.g., `TextMessage`, `StreamMessage`), a custom message property, or even the `JMSType` message header property. Since the `JMSType` header property may be used by the JMS provider, it is not a good idea to use it to store your own custom routing information or message type. In the following example, a single queue (`requestQueue`) is used to handle book orders, order status requests, and order cancellations. A custom message property is used to store the message type that is used to route the message to the particular class responsible for processing that message:

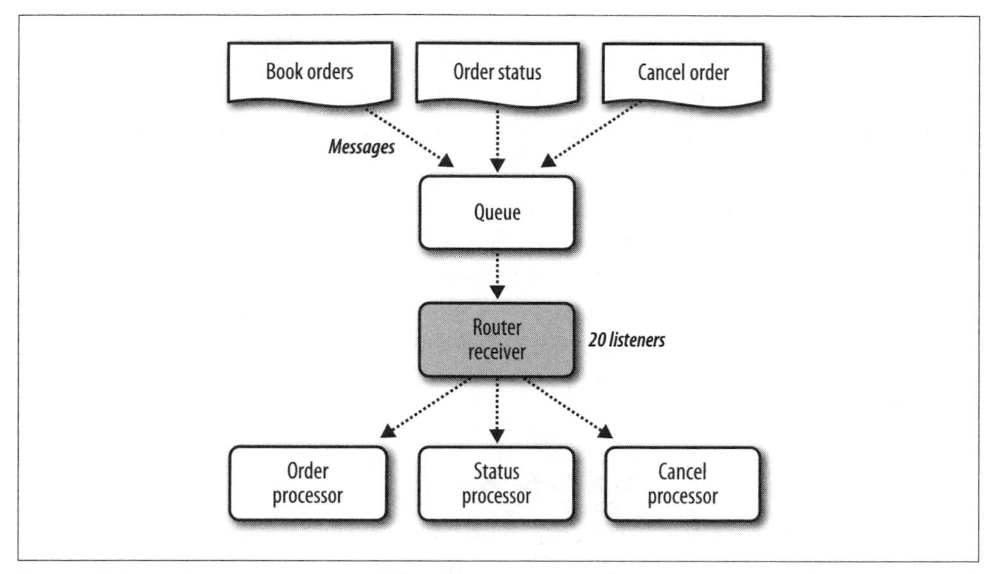

Figure 11-3. Single-purpose queue—different message types

```
public class QRouter implements MessageListener {

   private OrderProcessor orderProcessor = null;
   private StatusProcessor statusProcessor = null;
   private CancelProcessor cancelprocessor = null;

   ...
   public void onMessage(Message message) {
      try {
         //get the message payload
         String xml = ((TextMessage)message).getText();

         //get the message type
         int type = message.getIntProperty("type");
         if (type == NEW_BOOK_ORDER) {
            orderProcessor.placeOrder(xml);
         } else if (type == ORDER_STATUS) {
            statusProcessor.checkOrderStatus(xml);
         } else if (type == CANCEL_ORDER) {
            cancelProcessor.cancelOrder(xml);
         } else {
            throw new Exception("Invalid Order Type: " + type);
         }
      }
     ...
   }
   ...
}
```

In this code snippet, the XML application data is extracted from the message payload, then the message type is extracted from the message properties. Notice that the message type comes from a custom application property (`type`), not a standard header property. The `QRouter` class then analyzes the `type` value and redirects processing to one of three processor classes.

At first glance, this messaging design strategy may seem attractive due to the simplicity of the message processing. Adding a new messaging type (e.g., `VIEW_ORDER_HISTORY`) is simply a matter of adding the additional `if` statement in the `QRouter` class and adding the class or method to process that request. Since there is only a single queue, no additional messaging configuration is necessary to add the additional request:

```
public void onMessage(Message message) {
   try {
      ...
      //get the message type
      int type = message.getIntProperty("type");
      if (type == NEW_BOOK_ORDER) {
         orderProcessor.placeOrder(xml);
      } else if (type == ORDER_STATUS) {
         statusProcessor.checkOrderStatus(xml);
      } else if (type == CANCEL_ORDER) {
         cancelProcessor.cancelOrder(xml);
      } else if (type == VIEW_ORDER_HISTORY) {
         orderProcessor.viewHistory(xml);
      } else {
         throw new Exception("Invalid Order Type: " + type);
      }
   }
  ...
}
```

While this design approach seems to provide a lot of flexibility to the architecture, several inefficiencies manifest themselves in this design. First, since all messages are sent to the same queue, it is not possible to load balance the system based on the message type. For instance, in the preceding example, assume that there are 20 concurrent `QRouter` listener threads. This means 20 messages can be processed at the same time. However, let's say the distribution of message types is 80% `NEW_BOOK_ORDER`, 10% `ORDER_STATUS`, 7% `CANCEL_ORDER`, and 3% `VIEW_ORDER_HISTORY`. Therefore, if the queue contains 100 new book orders and a message is sent to cancel an order, the `CANCEL_ORDER` message will be placed on the queue at position 101 and not received until the previous 100 book orders are processed.

The issue is that the `QRouter` listener simply pulls off the next available message in the queue, regardless of the type. With this design it is not possible to tune the system to provide optimized throughput for the different message types. Using the original example, a better design approach would be to have three separate queues to handle the three message types and three separate message listeners, one for each queue. Figure 11-4 illustrates this approach.

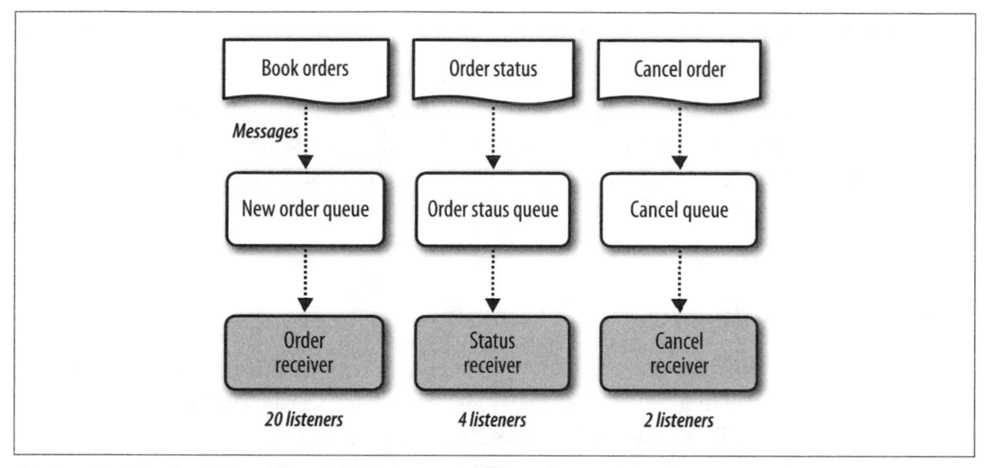

Figure 11-4. Multiple queues—different message types

Notice in Figure 11-4 how each message receiver can now be tuned to add the number of concurrent listener threads based on the message distribution. More importantly however, with this design 20 new book orders can still be processed concurrently, but order status requests and cancel order requests can immediately be processed at the same time. This is a significant improvement over the previous single-purpose queue design. With the multiple queue approach, adding new message types involves adding a new queue (configuration), changing the message producer code to send the new request to the new queue, and adding the new message listener. Since there is no central "routing listener," no additional coding changes are required.

What about a single queue that handles the same message type? In most cases this is exactly what we want, but in some cases this too can present issues. Take, for example, the book order messages from the previous example. After applying the multiple queue approach, we now have a queue dedicated to book orders. However, suppose that book orders can come from an online web-based source but also in batch form from a book store. They are the same `NEW_BOOK_ORDER` message, but they come from different client channels (one online, one batch). Now suppose the online web-based channel requires immediate feedback about the book order, whereas the batch orders from the book store (typically hundreds of book orders in a single batch) do not require a response. This is where additional issues can arise.

Just because the message type (e.g., `NEW_BOOK_ORDER`) is the same does not necessarily mean the messages themselves are the same. In the previous example, even though the message format and structure are exactly the same, there are actually *two types* of `NEW_BOOK_ORDER` messages: online and batch. When this scenario occurs, the same problems described with the multiple message types can occur here as well. Consider for a moment the case where a batch order comes in for 400 books. The queue depth for the *new order queue* is now 400. Immediately after the batch messages are received, an online request is received for a `NEW_BOOK_ORDER`. The online order is placed at position

401 in the new order queue, and will not be processed for quite some time. Meanwhile, the online web-based customer is waiting for the response for the book order.

A temporary solution to this problem is to use *message priority* to assign a higher priority to online messages, therefore effectively moving online messages to the front of the queue. However, using this approach leads to another messaging anti-pattern known as *message priority overuse*. This messaging anti-pattern is described in the following section.

Message Priority Overuse

In the previous section, we used message priority as a way to solve the problem where the same message type (e.g., `NEW_BOOK_ORDER`) is being used in two different ways. In the previous example, we had new book orders coming from an online web-based channel as well as a batch channel. Using message priority effectively moved the online orders ahead of the batch orders, solving the long wait problem for online book orders—or did it?

While the use of message priority as a way of processing certain messages faster may seem like a good permanent solution, the problem is that all of the listener threads available to process the higher priority messages may be tied up processing lower priority messages. Therefore, priority messages are processed slower than they otherwise could be.

In the example used in the previous section, online orders were given a higher priority than batch orders, meaning that they were pushed to the front of the queue. Suppose that, because of special validation and processing, batch orders take one minute to process, whereas online orders take 200 milliseconds to process. In this scenario most of the available message listeners will be tied up processing batch orders, leaving the online orders waiting on the queue to be processed. Once again, this is a case where the new order queue should be split into two queues: an online new order queue and a batch new order queue. This way, more listener threads can be assigned to the online new order queue where an immediate response is needed.

There are times when setting the message priority makes sense. However, a general rule of thumb to apply is as follows: when using message priority, always ask yourself if it would make more sense to use separate queues instead. Too often message priority is used to mask deeper rooted problems.

Message Header Misuse

Most of the message header properties are set by the JMS provider, even though the header property setter methods are exposed to the developer through the JMS API. This creates potential hard-to-find bugs, particularly when setting the message expiration and message priority.

The expiration date of a message is contained in the JMSExpiration message header with corresponding getJMSExpiration and setJMSExpiration methods to get and set the expiration date. All too often, a developer will attempt to set the expiration date of a message as follows:

```
public class QBorrower {
   ...
   private void sendLoanRequest(double salary, double loanAmt) {
      try {
         // Create JMS message
         MapMessage msg = qSession.createMapMessage();
         msg.setDouble("Salary", salary);
         msg.setDouble("LoanAmount", loanAmt);
         msg.setJMSReplyTo(responseQ);

         //set the message expiration to 30 seconds (incorrect!)
         msg.setJMSExpiration(new Date().getTime() + 30000);

         // Create the sender and send the message
         QueueSender qSender = qSession.createSender(requestQ);
         qSender.send(msg);
      ...
      }
   }
   ...
}
```

Notice the use of the setJMSExpiration method on the message object. This compiles fine and, when executed, will put the message on the queue. The message will then expire after 30 seconds if not received, right? Wrong. Using the code just shown, the message sent to the queue will never expire. Why? Because the time to live property on the message is set to zero (the default value). When the message is sent, the JMS provider adds the value in the time to live property to the current system time and sets the JMSExpiration header property itself. In the previous code example, the JMSExpiration header property was in fact getting overridden by the JMS provider.

This is a very common mistake and, unfortunately, is something that is rarely tested. So, if the setJMSExpiration method is off-limits to application developers, what is the proper way to set the message expiration? There are two ways to set the message expiration. The first technique is to invoke the setTimeToLive() method on the MessageProducer (QueueSender or TopicPublisher), which sets the message expiration for all messages sent using that sender:

```
//set the default message expiration for all messages to 30 seconds
QueueSender qSender = qSession.createSender(requestQ);
qSender.setTimeToLive(30000);
...
qSender.send(msg);
...
```

The other technique is to set the message expiration when sending the message:

```
QueueSender qSender = qSession.createSender(requestQ);
...
//set the message expiration for this message to 20 seconds
qSender.send(msg, DeliveryMode.PERSISTENT, 4, 20000);
...
```

Notice in this code snippet that messages sent using the `qSender` will, *by default*, have no message expiration. However, the message being sent in the code snippet has a message expiration of 20 seconds. Unfortunately, when using the second approach, you have to specify the message delivery mode and message priority as well.

There may be cases when you want to have a default message expiration for all messages sent by a `QueueSender` but still have the flexibility to override it for certain message types sent by that same `QueueSender`. In this case, you can use both forms together:

```
//set the default message expiration for all messages to 30 seconds
QueueSender qSender = qSession.createSender(requestQ);
qSender.setTimeToLive(30000);
...
qSender.send(msg1);

//this message should expire in 20 seconds
qSender.send(msg2, DeliveryMode.PERSISTENT, 0, 20000);
...
```

In this example, `msg1` will go on the queue and expire in 30 seconds (the default), whereas `msg2` will go on the same queue but expire in 20 seconds.

The same problem holds true for message priority, which is even harder to test than the message expiration. Here, a common mistake is to set the message priority directly using the `setJMSPriority()` method on the `Message` object prior to sending the message:

```
public class QBorrower {
   ...
   private void sendLoanRequest(double salary, double loanAmt) {
      try {
         // Create JMS message
         MapMessage msg = qSession.createMapMessage();
         msg.setDouble("Salary", salary);
         msg.setDouble("LoanAmount", loanAmt);

         //incorrect!
         msg.setJMSPriority(9);

         // Create the sender and send the message
         QueueSender qSender = qSession.createSender(requestQ);
         qSender.send(msg);
      ...
      }
   }
   ...
}
```

In this code, the message priority is set to 9, indicating this is a high-priority message. However, when the message is sent, the message will have a priority of 4 (normal priority). The reason? Like the message expiration, the JMS provider will look at the message priority property on the message and invoke the setJMSPriority method prior to placing the message on the queue. Since the default message priority is 4 (normal priority), the message priority will not be set to a high priority message, as the developer had originally intended.

Like the message expiration, there are two ways of setting the message priority: you can invoke the setPriority() method on the MessageProducer (QueueSender or TopicPublisher) or set the message priority when sending the message:

```
//set the default message priority for all messages to 9 (high)
QueueSender qSender = qSession.createSender(requestQ);
qSender.setPriority(9);
...
qSender.send(msg1);

//this message is low priority
qSender.send(msg2, DeliveryMode.PERSISTENT, 1, 30000);
...
```

In this example, msg1 will be sent with a priority of 9 (high priority), whereas msg2 will be sent with a priority of 1 (low priority).

Understanding these messaging anti-patterns will help you build more robust messaging systems and help you avoid some of the more common mistakes associated with messaging.

APPENDIX A

The Java Message Service API

This appendix is a quick reference guide to the Java Message Service API. It is organized into five sections: "Message Interfaces" (next), "Common Facilities" on page 249, "Common API" on page 252, "Point-to-Point API" on page 257, and "Publish-and-Subscribe API" on page 260. Each section provides a summary of its interfaces and is organized alphabetically. The XA-compliant interfaces are not included in this section because they are essentially the same as their non-XA interfaces. In addition, the Application Server API (`ConnectionConsumer`, `ServerSession`, and `ServerSessionPool`) is not covered in this book because this API is not supported by most vendors.

Message Interfaces

This section covers the message interface and the six message types.

BytesMessage

This `Message` type carries an array of primitive bytes as its payload. It's useful for exchanging data in an application's native format, providing for a high degree of interoperability with other messaging servers. It is also useful where JMS is used purely as a transport between two systems, and the message payload is opaque to the JMS client:

```
public interface BytesMessage extends Message {

    public long getBodyLength() throws JMSException
    public byte readByte() throws JMSException;
    public void writeByte(byte value) throws JMSException;
    public int readUnsignedByte() throws JMSException;
    public int readBytes(byte[] value) throws JMSException;
    public void writeBytes(byte[] value) throws JMSException;
    public int readBytes(byte[] value, int length)
        throws JMSException;
    public void writeBytes(byte[] value, int offset, int length)
        throws JMSException;
    public boolean readBoolean() throws JMSException;
    public void writeBoolean(boolean value) throws JMSException;
```

```
    public char readChar() throws JMSException;
    public void writeChar(char value) throws JMSException;
    public short readShort() throws JMSException;
    public void writeShort(short value) throws JMSException;
    public int readUnsignedShort() throws JMSException;
    public void writeInt(int value) throws JMSException;
    public int readInt() throws JMSException;
    public void writeLong(long value) throws JMSException;
    public long readLong() throws JMSException;
    public float readFloat() throws JMSException;
    public void writeFloat(float value) throws JMSException;
    public double readDouble() throws JMSException;
    public void writeDouble(double value) throws JMSException;
    public String readUTF() throws JMSException;
    public void writeUTF(String value) throws JMSException;
    public void writeObject(Object value) throws JMSException;
    public void reset() throws JMSException;
}
```

MapMessage

This `Message` type carries a set of name-value pairs as its payload. The payload is similar to a `java.util.Properties` object, except the values must be Java primitives or their wrappers. The `MapMessage` is useful for delivering keyed data:

```
public interface MapMessage extends Message {

    public boolean getBoolean(String name) throws JMSException;
    public void setBoolean(String name, boolean value)
        throws JMSException;
    public byte getByte(String name) throws JMSException;
    public void setByte(String name, byte value) throws JMSException;
    public byte[] getBytes(String name) throws JMSException;
    public void setBytes(String name, byte[] value)
        throws JMSException;
    public void setBytes(String name, byte[] value,
                         int offset, int length)
        throws  JMSException;
    public short getShort(String name) throws JMSException;
    public void setShort(String name, short value) throws JMSException;
    public char getChar(String name) throws JMSException;
    public void setChar(String name, char value) throws JMSException;
    public int getInt(String name) throws JMSException;
    public void setInt(String name, int value)throws JMSException;
    public long getLong(String name) throws JMSException;
    public void setLong(String name, long value) throws JMSException;
    public float getFloat(String name) throws JMSException;
    public void setFloat(String name, float value)
        throws JMSException;
    public double getDouble(String name) throws JMSException;
    public void setDouble(String name, double value)
        throws JMSException;
    public String getString(String name) throws JMSException;
    public void setString(String name, String value)
```

```
        throws JMSException;
    public Object getObject(String name) throws JMSException;
    public void setObject(String name, Object value)
        throws JMSException;
    public Enumeration getMapNames( ) throws JMSException;
    public boolean itemExists(String name) throws JMSException;
}
```

Message

The `Message` interface is the super interface for all message types. There are six messages types, including: `Message`, `TextMessage`, `ObjectMessage`, `StreamMessage`, `BytesMessage`, and `MapMessage`. The `Message` type has no payload and can be used for simple event notification.

A message basically has two parts: a *header* and a *payload*. The header is comprised of special fields that are used to identify the message, declare attributes of the message, and provide information for routing. The difference between message types is determined largely by their payload, which determines the type of application data the message contains:

```
public interface Message {
    public void acknowledge() throws JMSException;
    public void clearBody() throws JMSException;

    public Destination getJMSDestination() throws JMSException;
    public void setJMSDestination(Destination destination)
        throws JMSException;
    public int getJMSDeliveryMode() throws JMSException;
    public void setJMSDeliveryMode(int deliveryMode)
        throws JMSException;
    public String getJMSMessageID() throws JMSException;
    public void setJMSMessageID(String id) throws JMSException;
    public long getJMSTimestamp() throws JMSException;
    public void setJMSTimestamp(long timestamp) throws JMSException
    public long getJMSExpiration() throws JMSException;
    public void setJMSExpiration(long expiration) throws JMSException;
    public boolean getJMSRedelivered() throws JMSException;
    public void setJMSRedelivered(boolean redelivered)
        throws JMSException;
    public int getJMSPriority() throws JMSException;
    public void setJMSPriority(int priority) throws JMSException;
    public Destination getJMSReplyTo() throws JMSException;
    public void setJMSReplyTo(Destination replyTo) throws JMSException;
    public String getJMSCorrelationID() throws JMSException;
    public void setJMSCorrelationID(String correlationID)
        throws JMSException;
    public byte[] getJMSCorrelationIDAsBytes() throws JMSException;
    public void setJMSCorrelationIDAsBytes(byte[] correlationID)
        throws JMSException;
    public String getJMSType() throws JMSException;
    public void setJMSType(String type) throws JMSException;
```

```
    public String getStringProperty(String name)
        throws JMSException, MessageFormatException;
    public void setStringProperty(String name, String value)
        throws JMSException, MessageNotWriteableException;
    public int getIntProperty(String name)
        throws JMSException, MessageFormatException;
    public void setIntProperty(String name, int value)
        throws JMSException, MessageNotWriteableException;
    public boolean getBooleanProperty(String name)
        throws JMSException, MessageFormatException;
    public void setBooleanProperty(String name, boolean value)
        throws JMSException, MessageNotWriteableException;
    public double getDoubleProperty(String name)
        throws JMSException, MessageFormatException;
    public void setDoubleProperty(String name, double value)
        throws JMSException, MessageNotWriteableException;
    public float getFloatProperty(String name)
        throws JMSException, MessageFormatException;
    public void setFloatProperty(String name, float value)
        throws JMSException, MessageNotWriteableException;
    public byte getByteProperty(String name)
        throws JMSException, MessageFormatException;
    public void setByteProperty(String name, byte value)
        throws JMSException, MessageNotWriteableException;
    public long getLongProperty(String name)
        throws JMSException, MessageFormatException;
    public void setLongPreperty(String name, long value)
        throws JMSException, MessageNotWriteableException;
    public short getShortProperty(String name)
        throws JMSException, MessageFormatException;
    public void setShortProperty(String name, short value)
        throws JMSException, MessageNotWriteableException;
    public Object getObjectProperty(String name)
        throws JMSException, MessageFormatException;
    public void setObjectProperty(String name, Object value)
        throws JMSException, MessageNotWriteableException;
    public void clearProperties()
        throws JMSException;
    public Enumeration getPropertyNames()
        throws JMSException;
    public boolean propertyExists(String name)
        throws JMSException;
}
```

ObjectMessage

This Message type carries a serializable Java object as its payload. It is useful for exchanging Java objects:

```
public interface ObjectMessage extends Message {
    public java.io.Serializable getObject()
        throws JMSException;
    public void setObject(java.io.Serializable payload)
        throws JMSException, MessageNotWriteableException;
}
```

StreamMessage

This Message type carries a stream of primitive Java types (int, double, char, etc.) as its payload. It provides a set of convenience methods for mapping a formatted stream of bytes to Java primitives. It provides an easy programming model for exchanging primitive application data in a fixed order:

```
public interface StreamMessage extends Message {

    public boolean readBoolean() throws JMSException;
    public void writeBoolean(boolean value) throws JMSException;
    public byte readByte() throws JMSException;
    public int readBytes(byte[] value) throws JMSException;
    public void  writeByte(byte value) throws JMSException;
    public void writeBytes(byte[] value) throws JMSException;
    public void writeBytes(byte[] value, int offset, int length)
        throws JMSException;
    public short readShort() throws JMSException;
    public void writeShort(short value) throws JMSException;
    public char readChar() throws JMSException;
    public void writeChar(char value) throws JMSException;
    public int readInt() throws JMSException;
    public void writeInt(int value) throws JMSException;
    public long readLong() throws JMSException;
    public void writeLong(long value) throws JMSException;
    public float readFloat() throws JMSException;
    public void writeFloat(float value) throws JMSException;
    public double readDouble() throws JMSException;
    public void  writeDouble(double value) throws JMSException;
    public String  readString() throws JMSException;
    public void writeString(String value) throws JMSException;
    public Object readObject() throws JMSException;
    public void writeObject(Object value) throws JMSException;
    public void reset() throws JMSException;
}
```

TextMessage

This Message type carries a java.lang.String as its payload. It is useful for exchanging simple text messages and for more complex character data, such as XML documents:

```
public interface TextMessage extends Message {
    public String getText()
        throws JMSException;
    public void setText(String payload)
        throws JMSException, MessageNotWriteableException;
}
```

Common Facilities

This section covers additional messaging interfaces used by the common, point-to-point and publish-and-subscribe messaging APIs.

ConnectionMetaData

This type of object is obtained from a `Connection`, `TopicConnection`, or `QueueConnection` object. It provides information describing the JMS connection and the JMS provider. Information available includes the identity of the JMS provider, the JMS version supported by the provider, JMS provider version numbers, and the JMS properties supported:

```
public interface ConnectionMetaData {
    public int getJMSMajorVersion() throws JMSException;
    public int getJMSMinorVersion() throws JMSException;
    public String getJMSProviderName() throws JMSException;
    public String getJMSVersion() throws JMSException;
    public Enumeration getJMSXPropertyNames() throws JMSException;
    public int getProviderMajorVersion() throws JMSException;
    public int getProviderMinorVersion() throws JMSException;
    public String getProviderVersion() throws JMSException;
}
```

DeliveryMode

This class contains two final static variables, `PERSISTENT` and `NON_PERSISTENT`. These variables are used when establishing the delivery mode of a `MessageProducer`, `TopicPublisher`, or `QueueSender`.

There are two types of delivery modes in JMS: persistent and nonpersistent. A persistent message should be delivered *once-and-only-once*, which means that a message is not lost if the JMS provider fails; it will be delivered after the server recovers. A nonpersistent message is delivered *at-most-once*, which means that it can be lost and never delivered if the JMS provider fails. The default mode is `PERSISTENT`:

```
public interface DeliveryMode {
   public static final int NON_PERSISTENT = 1;
   public static final int PERSISTENT = 2;
}
```

ExceptionListener

JMS provides an `ExceptionListener` interface for trapping a lost connection and notifying the client of this condition. The `ExceptionListener` is bound to the connection. The `ExceptionListener` is very useful to JMS clients that wait passively for messages to be delivered and otherwise have no way of knowing that a connection has been lost.

It is the responsibility of the JMS provider to call the `onException()` method of all registered `ExceptionListeners` after making reasonable attempts to reestablish the connection automatically. The JMS client can implement the `ExceptionListener` so that it can be alerted to a lost connection, and possibly attempt to reestablish the connection manually:

```
public interface ExceptionListener {
   public void onException(JMSException exception);
}
```

JMSException

The `JMSException` is the base exception type for all exceptions thrown by the JMS API. It may provide an error message describing the cause of the exception, a provider-specific error code, and possibly a reference to the exception that caused the JMS exception:

```
public class JMSException extends java.lang.Exception {
    public JMSException(java.lang.String reason) { .. }
    public JMSException(java.lang.String reason,
                        java.lang.String errorCode) { .. }
    public String getErrorCode() { .. }
    public Exception getLinkedException() { .. }
    public void setLinkedException(java.lang.Exception ex) { .. }
}
```

While the `JMSException` is usually declared as the exception type thrown from methods in the JMS API, the actual exception thrown may be one of a dozen subtypes, which are enumerated here. The descriptions of these exception types are derived from Sun Microsystems' JMS API documentation, and they implement the methods defined by the `JMSException` super type:

`IllegalStateException`
: Thrown when a method is invoked illegally or inappropriately, or if the provider is not in an appropriate state when the method is called. For example, this exception should be thrown if `Session.commit()` is called on a nontransacted session.

`InvalidClientIDException`
: Thrown when a client attempts to set a connection's client ID to a value that the provider rejects.

`InvalidDestinationException`
: Thrown when the provider doesn't understand the destination, or the destination is no longer valid.

`InvalidSelectorException`
: Thrown when the syntax of a message selector is invalid.

`JMSSecurityException`
: Thrown when a provider rejects a username/password. Also thrown when a security restriction prevents a method from completing.

`MessageEOFException`
: Thrown if a stream ends unexpectedly when a `StreamMessage` or `BytesMessage` is being read.

`MessageFormatException`
: Thrown when a JMS client attempts to use a data type not supported by a message, or attempts to read data in a message as the wrong type. Also thrown when type errors are made with message property values.

`MessageNotReadableException`
: Thrown when a JMS client tries to read a write-only message.

`MessageNotWriteableException`
: Thrown when a JMS client tries to write to a read-only message.

`ResourceAllocationException`
: Thrown when a provider is unable to allocate the resources required by a method. This exception should be thrown when a call to `createTopicConnection()` fails because the JMS provider has insufficient resources.

`TransactionInProgressException`
: Thrown when an operation is invalid because a transaction is in progress. For instance, it should be thrown if you call `Session.commit()` when a session is part of a distributed transaction.

`TransactionRolledBackException`
: Thrown when calling `Session.commit()` results in a transaction rollback.

MessageListener

The `MessageListener` is implemented by the JMS client. It receives messages asynchronously from one or more `Consumers` (`TopicSubscriber` or `QueueReceiver`).

The `Session` (`TopicSession` or `QueueSession`) must ensure that messages are passed to the `MessageListener` serially, so that the messages can be processed separately. A `MessageListener` object may be registered with many consumers, but serial delivery is only guaranteed if all of its consumers were created by the same session:

```
public interface MessageListener {
    public void onMessage(Message message);
}
```

Common API

This section covers the common API new to JMS 1.1 that bridges the point-to-point and publish-and-subscribe APIs. The common API interfaces can be used with both queues and topics.

Connection

The `Connection` is the base interface for the `TopicConnection` and the `QueueConnection`, and represents an open TCP/IP socket to the JMS provider. It defines several general-purpose methods used by clients of the messaging system in managing a JMS connection. Among these methods are the getMetaData(), start(), stop(), and close() methods:

```
public interface Connection {

    public Session createSession(boolean transacted,
          int acknowledgeMode) throws JMSException;

    public ExceptionListener getExceptionListener() throws JMSException;
    public void setExceptionListener(ExceptionListener listener)
       throws JMSException;

    public ConnectionMetaData getMetaData() throws JMSException;

    public String getClientID() throws JMSException;
    public void setClientID(String clientID) throws JMSException;

    public void start() throws JMSException;
    public void stop() throws JMSException;
    public void close() throws JMSException;

    public ConnectionConsumer createConnectionConsumer
                                  (Destination destination,
                                   String messageSelector,
                                   ServerSessionPool sessionPool,
                                   int maxMessages)
       throws JMSException;
}
```

A Connection object represents a physical connection to a JMS provider for either point-to-point (QueueConnection) or publish-and-subscribe (TopicConnection) messaging. A JMS client might choose to create multiple connections from the same connection factory, but this is rare as connections are relatively expensive (each connection requires a network socket, I/O streams, memory, etc.). Creating multiple Session objects from the same Connection object is considered more efficient, because sessions share access to the same connection.

ConnectionFactory

The ConnectionFactory is the base type for the TopicConnectionFactory and the Queue ConnectionFactory, which are used in the publish-and-subscribe and point-to-point messaging models, respectively. It can also be used to create a Connection object.

The ConnectionFactory is implemented differently by each vendor, so configuration options available vary from product to product. A connection factory might, for example, be configured to manufacture connections that use a particular protocol, security scheme, clustering strategy, etc.:

```
public interface ConnectionFactory {
   public Connection createConnection() throws JMSException;
   public Connection createConnection(String userName,
                                      String password)
```

```
        throws JMSException
}
```

Destination

This interface is the base interface for the `Topic` and `Queue` interfaces, which represent destinations in the pub/sub and p2p domains, respectively.

In all modern enterprise messaging systems, applications exchange messages through virtual channels called *destinations*. When sending a message, the message is addressed to a destination, not a specific application. Any application that subscribes or registers an interest in that destination may receive that message. In this way, the applications that receive messages and those that send messages are decoupled. Senders and receivers are not bound to each other in any way and may send and receive messages as they see fit:

```
public interface Destination {
}
```

MessageConsumer

The `MessageConsumer` is the base interface for the `TopicSubscriber` and the `QueueReceiver`. It defines several general-purpose methods used by clients when using a consumer. Among these methods are the `setMessageListener()` and `close()` methods and three types of `receive()` methods.

`MessageConsumer` can consume messages asynchronously or synchronously. To consume messages asynchronously, the JMS client must provide the `MessageConsumer` with a `MessageListener` object, which will then receive the messages as they arrive. To consume messages synchronously, the JMS client may call one of three receive methods—`receive()`, `receive(long timeout)`, and `receiveNoWait()`:

```
public interface MessageConsumer {
    public void close() throws JMSException;
    public MessageListener getMessageListener() throws JMSException;
    public String getMessageSelector() throws JMSException;
    public Message receive() throws JMSException;
    public Message receive(long timeout) throws JMSException;
    public Message receiveNoWait() throws JMSException;
    public void setMessageListener(MessageListener listener)
        throws JMSException;
}
```

MessageProducer

The `MessageProducer` is the base interface for the `TopicPublisher` and the `QueueSender`. It defines several general-purpose methods used by clients. Among these methods are `setDeliveryMode()`, `close()`, `setPriority()`, and `setTimeToLive(long timeToLive)`.

MessageProducer sends messages to a specified destination (Topic or Queue). The default destination can be determined when the MessageProducer is created by its session, or the destination can be set each time a message is sent—in this case, there is no default destination:

```
public interface MessageProducer {
    public void setDisableMessageID(boolean value) throws JMSException;
    public boolean getDisableMessageID() throws JMSException;
    public void setDisableMessageTimestamp(boolean value)
        throws JMSException;
    public boolean getDisableMessageTimestamp() throws JMSException;
    public void setPriority(int defaultPriority) int getDeliveryMode()
        throws JMSException;
    public int getPriority() throws JMSException;
    public void setTimeToLive(long timeToLive) throws JMSException;
    public long getTimeToLive() throws JMSException;
    public void close() throws JMSException;

    public void send(Message message)
        throws JMSException;
    public void send(Destination destination,
                     Message message)
        throws JMSException;
    public void send(Message message,
                     int deliveryMode,
                     int priority,
                     long timeToLive)
        throws JMSException;
    public void send(Destination destination,
                     Message message,
                     int deliveryMode,
                     int priority,
                     long timeToLive)
        throws JMSException

    public Destination getDestination() throws JMSException;
    public int getDeliveryMode() throws JMSException;
    public void setDeliveryMode(int deliveryMode) throws JMSException;
}
```

Session

The Session is the base interface for the TopicSession and the QueueSession. It defines several general-purpose methods used by JMS clients for managing a JMS Session. Among these methods are the six createMessage() methods (one for each type of Message object), setMessageListener(), close(), and transaction methods.

A Session is a single-threaded context for producing and consuming messages. It creates message consumers, producers, and messages for a specific JMS provider. The Session manages the scope of transactions across send and receive operations, tracks message acknowledgment for consumers, and serializes delivery of messages to MessageListener objects:

```
public interface Session extends java.lang.Runnable {
    public static final int AUTO_ACKNOWLEDGE = 1;
    public static final int CLIENT_ACKNOWLEDGE = 2;
    public static final int DUPS_OK_ACKNOWLEDGE = 3;
    public static final int SESSION_TRANSACTED = 4;

    public BytesMessage createBytesMessage() throws JMSException;
    public MapMessage createMapMessage() throws JMSException;
    public Message createMessage() throws JMSException;
    public ObjectMessage createObjectMessage() throws JMSException;
    public ObjectMessage createObjectMessage(Serializable object)
        throws JMSException;
    public StreamMessage createStreamMessage() throws JMSException;
    public TextMessage createTextMessage() throws JMSException;
    public TextMessage createTextMessage(String text)
        throws JMSException;

    public boolean getTransacted() throws JMSException;
    public int getAcknowledgeMode() throws JMSException;
    public void unsubscribe(String name) throws JMSException;
    public void commit() throws JMSException;
    public void rollback() throws JMSException;
    public void close() throws JMSException;
    public void recover() throws JMSException;
    public void run();

    public MessageListener getMessageListener() throws JMSException;
    public void setMessageListener(MessageListener listener)
        throws JMSException;

    public QueueBrowser createBrowser(Queue queue)
        throws JMSException;
    public QueueBrowser createBrowser(Queue queue,
                                      String messageSelector)
        throws JMSException;

    public MessageConsumer createConsumer(Destination destination)
        throws JMSException;
    public MessageConsumer createConsumer(Destination destination,
                                          String messageSelector)
        throws JMSException;
    public MessageConsumer createConsumer(Destination destination,
                                          String messageSelector,
                                          boolean NoLocal)
        throws JMSException;
    public MessageProducer createProducer(Destination destination)
                                          throws JMSException;

    public TopicSubscriber createDurableSubscriber(Topic topic,
                                                   String name)
        throws JMSException;
    public TopicSubscriber createDurableSubscriber(Topic topic,
                                                   String name,
                                                   String messageSelector,
                                                   boolean noLocal)
```

```
        throws JMSException;

    public TemporaryQueue createTemporaryQueue()
        throws JMSException;
    public TemporaryTopic createTemporaryTopic()
        throws JMSException;

    public Queue createQueue(String queueName)
        throws JMSException;
    public Topic createTopic(String topicName)
        throws JMSException;
}
```

Point-to-Point API

This section covers the queue-based point-to-point interfaces and classes.

Queue

The `Queue` is an administered object that acts as a handle or identifier for an actual queue, called a *physical queue* , on the messaging server. A physical queue is a channel through which many clients can receive and send messages. The `Queue` is a subtype of the `Destination` interface.

Multiple receivers may connect to a queue, but each message in the queue may only be consumed by one of the queue's receivers. Messages in the queue are ordered so that consumers receive messages in the order the message server placed them in the queue:

```
public interface Queue extends Destination {
    public String getQueueName() throws JMSException;
    public String toString();
}
```

QueueBrowser

A `QueueBrowser` is a specialized object that allows you to peek ahead at pending messages on a `Queue` without actually consuming them. This feature is unique to point-to-point messaging. Queue browsing can be useful for monitoring the contents of a queue from an administration tool, or for browsing through multiple messages to locate a message that is more important than the one that is at the head of the queue:

```
public interface QueueBrowser {
    public Queue getQueue() throws JMSException;
    public String getMessageSelector() throws JMSException;
    public Enumeration getEnumeration() throws JMSException;
    public void close() throws JMSException;
}
```

QueueConnection

The `QueueConnection` is created by the `QueueConnectionFactory`. Each `QueueConnection` represents a unique connection to the server.[*] The `QueueConnection` is a subtype of the `Connection` interface:

```
public interface QueueConnection extends Connection {
    public QueueSession createQueueSession(boolean transacted,
                                           int acknowledgeMode)
        throws JMSException;
    public ConnectionConsumer createConnectionConsumer
                         (Queue queue,
                          String messageSelector,
                          ServerSessionPool sessionPool,
                          int maxMessages)
        throws JMSException;
}
```

QueueConnectionFactory

The `QueueConnectionFactory` is an administered object that is used to manufacture `QueueConnectionFactory` objects. The `QueueConnectionFactory` is a subtype of the `ConnectionFactory` interface:

```
public interface QueueConnectionFactory extends ConnectionFactory {
    public QueueConnection createQueueConnection() throws JMSException;
    public QueueConnection createQueueConnection(String username, String password)
        throws JMSException;
}
```

QueueReceiver

The `QueueReceiver` is created by a `QueueSession` for a specific queue. The JMS client uses the `QueueReceiver` to receive messages delivered to its assigned queue. The `QueueReceiver` is a subtype of the `MessageConsumer` interface.

Each message in a queue is delivered to only one `QueueReceiver`. Multiple receivers may connect to a queue, but each message in the queue may only be consumed by one of the queue's receivers:

```
public interface QueueReceiver extends MessageConsumer {
    public Queue getQueue() throws JMSException;
}
```

[*] The actual physical network connection may or may not be unique, depending on the vendor. However, the connection is considered to be logically unique so authentication and connection control can be managed separately from other connections.

QueueRequestor

The QueueRequestor class is used for request/reply processing. It creates a temporary queue to be used for the response message; therefore, no message correlation is needed:

```
public class QueueRequestor extends Object {

    public QueueRequestor(QueueSession session,
                          Queue queue)
       throws JMSException;

    public Message request(Message message)
       throws JMSException;
    public void close() throws JMSException;
}
```

QueueSender

A QueueSender is created by a QueueSession, usually for a specific queue. Messages sent by the QueueSender to a queue are delivered to a client connected to that queue. The QueueSender is a subtype of the MessageProducer interface:

```
public interface QueueSender extends MessageProducer {
    public Queue getQueue() throws JMSException;
    public void send(Message message) throws JMSException;
    public void send(Message message, int deliveryMode, int priority,
                     long timeToLive)
        throws JMSException;
    public void send(Queue queue, Message message) throws JMSException;
    public void send(Queue queue, Message message,int deliveryMode,
                     int priority,long timeToLive)
        throws JMSException;
}
```

QueueSession

The QueueSession is created by the QueueConnection. A QueueSession object is a factory for creating Message, QueueSender, and QueueReceiver objects. A client can create multiple QueueSession objects to provide more granular control over senders, receivers, and their associated transactions. The QueueSession is a subtype of the Session interface:

```
public interface QueueSession extends Session {
    public Queue createQueue(java.lang.String queueName)
        throws JMSException;
    public QueueReceiver createReceiver(Queue queue)
        throws JMSException;
    public QueueReceiver createReceiver(Queue queue, String messageSelector)
        throws JMSException;
    public QueueSender createSender(Queue queue) throws JMSException;
    public QueueBrowser createBrowser(Queue queue) throws JMSException;
    public QueueBrowser createBrowser(Queue queue, String messageSelector)
        throws JMSException;
```

```
    public TemporaryQueue createTemporaryQueue() throws JMSException;
}
```

TemporaryQueue

A TemporaryQueue is created by a QueueSession or Session. A temporary queue is associated with the connection that belongs to the QueueSession that created it. It is only active for the duration of the session's connection, and is guaranteed to be unique across all connections. It lasts only as long as its associated client connection is active. In all other respects, a temporary queue is just like a "regular" queue. The TemporaryQueue is a subtype of the Queue interface.

Since a temporary queue is created by a JMS client, it is unavailable to other JMS clients unless the queue identity is transferred using the JMSReplyTo header. While any client may send messages on another client's temporary queue, only the sessions that are associated with the JMS client connection that created the temporary queue may receive messages from it. JMS clients can also, of course, send messages to their own temporary queues:

```
public interface TemporaryQueue extends Queue {
    public void delete() throws JMSException;
}
```

Publish-and-Subscribe API

This section covers the topic-based publish-and-subscribe interfaces and classes.

TemporaryTopic

A TemporaryTopic is created by a TopicSession or Session. A temporary topic is associated with the connection that belongs to the TopicSession that created it. It is only active for the duration of the session's connection, and it is guaranteed to be unique across all connections. Since it is temporary it can't be durable—it lasts only as long as its associated client connection is active. In all other respects it is just like a "regular" topic. The TemporaryTopic is a subtype of the Topic interface.

Since a temporary topic is created by a JMS client, it is unavailable to other JMS clients unless the topic identity is transferred using the JMSReplyTo header. While any client may publish messages on another client's temporary topic, only the sessions that are associated with the JMS client connection that created the temporary topic may subscribe to it. JMS clients can also, of course, publish messages to their own temporary topics:

```
public interface TemporaryTopic extends Topic {
    public void delete() throws JMSException;
}
```

Topic

The `Topic` is an administered object that acts as a handle or identifier for an actual topic, called a *physical topic*, on the messaging server. A physical topic is a channel to which many clients can subscribe and publish. When a JMS client delivers a `Message` object to a topic, all the clients subscribed to that topic receive the `Message`. The `Topic` is a subtype of the `Destination` interface:

```
public interface Topic extends Destination {
    public String getTopicName() throws JMSException;
    public String toString();
}
```

TopicConnection

The `TopicConnection` is created by the `TopicConnectionFactory`. The `TopicConnection` represents a connection to the message server. Each `TopicConnection` created from a `TopicConnectionFactory` is a unique connection to the server.[†] The `TopicConnection` is a subtype of the `Connection` interface:

```
public interface TopicConnection extends Connection {
    public TopicSession createTopicSession(boolean transacted,
                                           int acknowledgeMode)
        throws JMSException;
    public ConnectionConsumer createConnectionConsumer
                               (Topic topic, String messageSelector,
                                ServerSessionPool sessionPool,
                                int maxMessages)
        throws JMSException;
    public ConnectionConsumer createDurableConnectionConsumer
                               (Topic topic, String subscriptionName,
                                String messageSelector,
                                ServerSessionPool sessionPool,
                                int maxMessages)
        throws JMSException;
}
```

TopicConnectionFactory

The `TopicConnectionFactory` is an administered object that is used to manufacture `TopicConnection` objects. The `TopicConnectionFactory` is a subtype of the `Connection Factory` interface:

```
public interface TopicConnectionFactory extends ConnectionFactory {
    public TopicConnection createTopicConnection() throws JMSException;
    public TopicConnection createTopicConnection(String username,
                                                 String password)
```

[†] The actual physical network connection may or may not be unique, depending on the vendor. However, the connection is considered to be logically unique so authentication and connection control can be managed separately from other connections.

```
        throws JMSException;
}
```

TopicPublisher

A TopicPublisher is created by a TopicSession, usually for a specific Topic. Messages that are sent by the TopicPublisher are copied and delivered to each client subscribed to that topic. The TopicPublisher is a subtype of the MessageProducer interface:

```
public interface TopicPublisher extends MessageProducer {
    public Topic getTopic() throws JMSException;
    public void publish(Message message) throws JMSException;
    public void publish(Message message, int deliveryMode,int priority,
                        long timeToLive)
        throws JMSException;
    public void publish(Topic topic,Message message)
        throws JMSException;
    public void publish(Topic topic, Message message, int deliveryMode,
                        int priority,long timeToLive)
        throws JMSException;
}
```

TopicRequestor

The TopicRequestor class is used for request/reply processing. It creates a temporary topic to be used for the response message; therefore, no message correlation is needed:

```
public class TopicRequestor extends Object {

    public TopicRequestor(TopicSession session,
                          Topic topic)
       throws JMSException;

    public Message request(Message message)
       throws JMSException;
    public void close() throws JMSException;
}
```

TopicSession

The TopicSession is created by the TopicConnection. A TopicSession object is a factory for creating Message, TopicPublisher, and TopicSubscriber objects. A client can create multiple TopicSession objects to provide more granular control over publishers, subscribers, and their associated transactions. The TopicSession is a subtype of the Session interface:

```
public interface TopicSession extends Session {
    public Topic createTopic(java.lang.String topicName)
        throws JMSException;
    public TopicSubscriber createSubscriber(Topic topic)
        throws JMSException;
    public TopicSubscriber createSubscriber(Topic topic,
                                            String messageSelector,
                                            boolean noLocal)
        throws JMSException;
    public TopicSubscriber createDurableSubscriber(Topic topic,
                                                   String name)
        throws JMSException;
    public TopicSubscriber createDurableSubscriber
                                              (Topic topic,
                                               String name,
                                               String messageSelector,
                                               boolean noLocal)
        throws JMSException;
    public TopicPublisher createPublisher(Topic topic)
        throws JMSException;
    public TemporaryTopic createTemporaryTopic() throws JMSException;
    public void unsubscribe(java.lang.String name) throws JMSException;
}
```

TopicSubscriber

The `TopicSubscriber` is created by a `TopicSession` for a specific topic. The messages are delivered to the `TopicSubscriber` as they become available, avoiding the need to poll the topic for new messages. The `TopicSubscriber` is a subtype of the `MessageConsumer` interface:

```
public interface TopicSubscriber extends MessageConsumer
    public Topic getTopic() throws JMSException;
    public boolean getNoLocal() throws JMSException;
}
```

APPENDIX B

Message Headers

The message headers provide metadata describing who or what created the message, when it was created, how long its data is valid, etc. The headers also contain routing information that describes the destination of the message (topic or queue), how a message should be acknowledged, and a lot more.

The `Message` interface provides mutator ("set") methods for each of the JMS headers, but only the `JMSReplyTo`, `JMSCorrelationID`, and `JMSType` headers can be modified using these methods. Calls to the mutator methods for any of the other JMS headers will be ignored when the message is sent. According to the authors of the specification, the mutator methods were left in the `Message` interface for "general orthogonality"; to balance the accessor methods—a fairly strange but well-established justification.

The accessor ("get") methods always provide the JMS client with information about the JMS headers. However, some JMS headers (`JMSTimestamp`, `JMSRedelivered`, etc.) are not available until after the message is sent or even received.

JMSDestination

Purpose: Routing

Message objects are always sent to some kind of destination. In the pub/sub model, `Message` objects are delivered to a topic, identified by a `Topic` object. In Chapter 2, you learned that the destination of a `Message` object is established when the `TopicPublisher` is created:

```
Topic chatTopic = (Topic)ctx.lookup(topicName);
TopicPublisher publisher = session.createPublisher(chatTopic);
```

The `JMSDestination` header identifies the destination of a `Message` object using a `javax.jms.Destination` object. The `Destination` class is the superclass of both `Topic` (pub/sub) and `Queue` (p2p). The `JMSDestination` header is obtained using the `Message.getJMSDestination()` method.

Identifying the destination to which a message was delivered is valuable to JMS clients that consume messages from more than one topic or queue. `MessageListener` objects might, for example, listen to multiple consumers (`TopicSubscriber` or `QueueReceiver` types) so that they receive messages from more than one topic or queue. For example, the `Chat` client from Chapter 2 could be modified to subscribe to more than one chat topic at a time. In this scenario, the `onMessage()` method of the `MessageListener` would use the `JMSDestination` header to identify which chat topic a message came from:

```
public void onMessage(Message message){
   try {
      TextMessage textMessage = (TextMessage)message;
      String text = textMessage.getText();

      Topic topic = (Topic)textMessage.getJMSDestination();
      System.out.println(topic.getTopicName()+": "+text);
   } catch (JMSException jmse){jmse.printStackTrace();}
}
```

The `JMSDestination` header is set automatically by the JMS provider when the message is delivered. The `Destination` used in the `JMSDestination` header is typically specified when the publisher is created, as shown here:

```
Queue queue = (Queue)ctx.lookup(queueName);
QueueSender queueSender = session.createSender(queue);
...
Topic topic = (Topic)ctx.lookup(topicName);
TopicPublisher topicPublisher = session.createPublisher(topic);
```

An unspecified message producer—one created without a `Destination`—will require that a `Destination` be supplied with each `send()` operation:

```
QueueSender queueSender = session.createSender(null);
Message message = session.createMessage();

Queue queue = (Queue)jndi.lookup(queueName);
queueSender.send(queue, message);
...
TopicPublisher topicPublisher = session.createPublisher(null);
Message message = session.createMessage();

Topic topic = (Topic)jndi.lookup(topicName);
topicPublisher.publish(topic, message);
```

In this case, the `JMSDestination` header becomes the `Destination` used in the `send()` operation.

JMSDeliveryMode

Purpose: Routing

There are two types of delivery modes in JMS: persistent and nonpersistent. A persistent message should be delivered once-and-only-once, which means that a message is not lost if the JMS provider fails; it will be delivered after the server recovers. A nonpersistent message is delivered at-most-once, which means that it can be lost and never delivered if the JMS provider fails. In both persistent and nonpersistent delivery modes the message server should not send a message to the same consumer more than once, but it is possible; see the section on `JMSRedelivered` for more details.

The vendor-supplied client runtime and the server functionality are collectively referred to as the JMS provider. A "provider failure" generically describes any failure condition that is outside of the domain of the application code. It could mean a hardware failure that occurs while the provider is entrusted with the processing of a message, or it could mean an unexpected exception or halting of a process due to a software defect. It could also mean a network failure that occurs between two processes that are part of the JMS vendor's internal architecture.

Persistent messages are intended to survive system failures of the JMS provider (the message server). Persistent messages are written to disk as soon as the message server receives them from the JMS client. After the message is persisted to disk the message server can then attempt to deliver the message to its intended consumer. As the messaging server delivers the message to the consumers it keeps track of which consumers successfully receive the message. If the JMS provider fails while delivering the message, the message server will pick up where it left off following a recovery. Persistent messages are delivered once-and-only-once. The mechanics of this are covered in greater detail in Chapter 7.

Nonpersistent messages are not written to disk when they are received by the message server, so if the JMS provider fails, the message will be lost. In general nonpersistent messages perform better than persistent messages. They are delivered more quickly and require less system resources on the message server. However, nonpersistent messages should only be used when a loss of messages due to a JMS provider failures is not an issue. The chat example used in Chapter 2 is a good example of a system that doesn't require persistent delivery. It's not critical that every message be delivered to all consumers in a chat application. In most business systems, however, messages are delivered using the persistent mode, because it's important that they be successfully delivered.

The delivery mode can be set using the `setDeliveryMode()` method defined in both the `TopicPublisher` and `QueueSender` message producers. The `javax.jms.DeliveryMode` class defines the two constants used to declare the delivery mode: `PERSISTENT` and `NON_PERSISTENT`:

```
// Publish-and-subscribe
TopicPublisher topicPublisher = topicSession.createPublisher(topic);
topicPublisher.setDeliveryMode(DeliveryMode.NON_PERSISTENT);

// Point-to-point
QueueSender queueSender = queueSession.createSender(queue);
queueSender.setDeliverMode(DeliveryMode.PERSISTENT);
```

Once the delivery mode has been set on the message producer, it will be applied to all the messages delivered by that producer. The delivery mode can be changed at any time using the `setDeliveryMode()` method; the new mode will be applied to subsequent messages. The default delivery mode of a message producer is always `PERSISTENT`.

The delivery mode of a message producer can be overridden for an individual message during the send operation, which allows a message producer to deliver a mixture of persistent and nonpersistent messages to the same destination (topic or queue):

```
// Publish-and-subscribe
Message message = topicSession.createMessage();
```

```
topicPublisher.publish(message, DeliveryMode.PERSISTENT, 5, 0);

// Point-to-point
Message message = queueSession.createMessage();
queueSender.send(message, DeliveryMode.NON_PERSISTENT, 5, 0);
```

The `JMSDeliveryMode` can be obtained from the `Message` object using the `getJMSDeliveryMode()` method:

```
public void onMessage(Message message){
    try {
     if (message.getJMSDeliveryMode() == DeliveryMode.PERSISTENT){
       // Do something
     } else {
       // Do something else
     }
    } catch (JMSException jmse){jmse.printStackTrace();}
}
```

JMSMessageID

Purpose: Routing

The `JMSMessageID` is a `String` value that uniquely identifies a message. How unique the identifier is depends on the vendor. It may only be unique for that installation of the message server, or it may be universally unique.

The `JMSMessageID` can be useful for historical repositories in applications where messages need to be uniquely indexed. The `JMSMessageID` is also useful for correlating messages, which is done using the `JMSCorrelationID` header. Use of the `JMSCorrelationID` header is described in more detail in Chapter 4.

The message provider generates the `JMSMessageID` automatically when the message is received from a JMS client. The `JMSMessageID` must start with `ID:`, but the rest of `JMSMessageID` can be any collection of characters that uniquely identifies the message to the JMS provider. Here is an example of a `JMSMessageID` generated by Progress's SonicMQ:

```
// JMSMessageID generated by SonicMQ
ID:6c867f96:20001:DF59525514
```

If a unique message ID is not needed by the JMS application, the JMS client can provide a hint to the message server that an ID is not necessary by using the `setDisableMessageID()` method (as shown in the following code). Vendors that heed this hint can reduce message processing time by not generating unique IDs for each message. If a `JMSMessageID` is not generated, the `getJMSMessageID()` method returns `null`:

```
// Publish-and-subscribe
TopicPublisher topicPublisher = topicSession.createPublisher(topic);
topicPublisher.setDisableMessageID(true);

// Point-to-point
QueueSender queueSender = queueSession.createSender(topic);
queueSender.setDisableMessageID(true);
```

JMSTimestamp

Purpose: Identification

The JMSTimestamp is set automatically by the message producer when the send operation is invoked. The value of the JMSTimestamp is the *approximate* time that the send operation was invoked. Sometimes messages are not transmitted to the message server immediately. A message can be delayed for many reasons, including how the message producer is configured, whether it's a transacted session, the acknowledgment mode used, etc. When the send() operation returns, the message object will have its timestamp:

```
Message message = topicSession.createMessage();
topicPublisher.publish(message);
long time = message.getJMSTimestamp();
```

The timestamp is set automatically, thus any value set explicitly by the JMS client will be ignored and discarded when the send() operation is invoked. The value of the timestamp is the amount of time, measured in milliseconds, that has elapsed since midnight, January 1, 1970, UTC (see "UTC" on page 271 for more information).

Timestamps can be used by message consumers as indicators of the approximate time that the message was delivered by the message producer. The timestamp can be useful when ordering messages or for historical repositories.

The JMSTimestamp is set during the send operation and may be calculated locally by the producer (TopicPublisher or QueueSender) on the client or it may be obtained from the message server. In the first case, when the producer calculates the timestamp, the timestamps can vary from JMS client to client. This is because the timestamp is obtained from the JMS client's local system clock, which may not be synchronized with other JMS client machines. Timestamps acquired from the message server are more consistent across JMS clients using the same JMS provider, since all the times are acquired from the same source, the common message server. It's possible to disable timestamps—or at least hint that they are not needed—by invoking the setDisableMessageTimestamp() method, available on both TopicPublisher and QueueSender objects:

```
// Publish-and-subscribe
TopicPublisher topicPublisher = topicSession.createPublisher(topic);
topicPublisher.setDisableMessageTimestamp(true);

// Point-to-point
QueueSender queueSender = queueSession.createSender(topic);
queueSender.setDisableMessageTimestamp(true);
```

If the JMS provider heeds the hint to disable the timestamp, the JMSTimestamp is set to 0, indicating that no timestamp was set. Disabling the timestamp can reduce the workload for JMS providers that use the message server to generate timestamps (instead of the JMS client), and can reduce the size of a message by at least 8 bytes (the size of a long value), which reduces the amount of network traffic. Support for disabling the timestamp is optional, which means that some vendors will set the timestamp whether you need it or not.

JMSExpiration

Purpose: Routing

A `Message` object can have an expiration date, the same as on a carton of milk. The expiration date is useful for messages that are only relevant for a fixed amount of time. The expiration time for messages is set in milliseconds by the producer using the `setTimeToLive()` method on either the `QueueSender` or `TopicPublisher` as shown below:

```
// Publish-and-subscribe
TopicPublisher topicPublisher = topicSession.createPublisher(topic);
// Set time to live as 1 hour (1000 millis x 60 sec x 60 min)
topicPublisher.setTimeToLive(3600000);

// Point-to-point
QueueSender queueSender = queueSession.createSender(topic);
// Set time to live as 2 days (1000 millis x 60 sec x 60 min x 48 hours)
queueSender.setTimeToLive(172800000);
```

By default the `timeToLive` is zero, which indicates that the message doesn't expire. Calling `setTimeToLive()` with a zero value as the argument ensures that message is created without an expiration date. The message expiration can also be set on the `send()` or `publish()` method of the message producer as well:

```
// Publish-and-subscribe
// Set time to live as 1 hour (1000 millis x 60 sec x 60 min)
Message message = topicSession.createMessage();
topicPublisher.publish(message, DeliveryMode.PERSISTENT, 5, 3600000);

// Point-to-point
// Set time to live as 2 days (1000 millis x 60 sec x 60 min x 48 hours)
Message message = queueSession.createMessage();
queueSender.send(message, DeliveryMode.NON_PERSISTENT, 5, 172800000);
```

The `JMSExpiration` date itself is calculated as:

```
JMSExpiration = currenttime + timeToLive.
```

The value of the `currenttime` is the amount of time, measured in milliseconds, that has elapsed since the Java epoch (midnight, January 1, 1970, UTC).

The JMS specification doesn't state whether the current time is calculated by the client computer or the message server, so consistency is dependent on either the accuracy of every client machine or the message server. We can certainly empathize with the JMS spec producers for remaining agnostic on this issue. Whether or not timestamps are synchronized across clients depends on the application. There is nothing preventing a JMS vendor from providing a configuration setting to control this behavior.

The `JMSExpiration` is the date and time that the message will expire. JMS clients should be written to discard any unprocessed messages that have expired, because the data and event communicated by the message is no longer valid. Message providers (servers) are also expected to discard any undelivered messages that expire while in their queues and topics. Even persistent messages are supposed to be discarded if they expire before being delivered.

UTC

UTC (Universal Time Coordinated, a.k.a. Coordinated Universal Time) is an internationally accepted official standard time based on the coordination of hundreds of atomic clocks worldwide. The JMS specification states that the time used to calculate the `JMSExpiration` and `JMSTimestamp` are based on UTC time, but in reality this is rarely the case. Ordinarily, there is a discrepancy between the current time reported by the Java Virtual Machine and the true UTC. This is because the system clocks on desktop computers and business servers are usually not synchronized with UTC, and are not accurate enough to keep UTC time. System clocks can be coordinated with the UTC through an Internet protocol called NTP (Network Time Protocol), which periodically queries for the actual UTC from a network time service and resynchronizes the system clock with the UTC.

You can get the system clock's time from any Java Virtual Machine using the `System` class as shown here:

```
long currentTime = System.currentTimeMillis();
```

The system clock's time, as reported by the JVM, is calculated as the number of milliseconds (1,000 milliseconds = 1 second) that have elapsed since January 1st, 1970, assuming that the system clock is reasonably accurate.

JMSRedelivered

Purpose: Routing

The `JMSRedelivered` header indicates if the message was redelivered to the consumer. The `JMSRedelivered` header is `true` if the message has been redelivered, and `false` if has not. A message may be marked as redelivered if a consumer failed to acknowledge delivery, or if the JMS provider is otherwise uncertain whether the consumer received the message.

When a message is delivered to a consumer, the consumer must acknowledge receipt of the message. If it doesn't, the message server may attempt to redeliver the message. Consumers can acknowledge messages automatically or manually, depending on how the consumer was created. A consumer created with an acknowledgment mode of `AUTO_ACKNOWLEDGE` or `DUPS_OK_ACKNOWLEDGE` automatically informs the message server that the message was received. When the consumer is created with `CLIENT_ACKNOWLEDGE` mode, the JMS client must manually acknowledge the messages using the `acknowledge()` method.

In general, when a message has a `JMSRedelivered` value of `false`, the consumer should assume that there is no chance it has seen this message before. If the redelivered flag is `true`, the client may have been given this message before so it may need to take some precautions it would not otherwise take. Redelivery can occur under a variety of conditions, and a JMS provider may mark a message as redelivered when it's in doubt due to failures, error conditions, and other anomalous conditions.

Message acknowledgment and redelivery are covered in detail in Chapter 7.

JMSPriority

Purpose: Routing

Messages may be assigned a priority by the message producer when they are delivered. The message servers may use message's priority to order delivery of messages to consumers; messages with a higher priority are delivered ahead of lower priority messages.

The message's priority is contained in the `JMSPriority` header, which is set automatically by the JMS provider. If not specified, the message priority is set to a default value of 4. The priority of messages can be declared by the JMS client using the `setPriority()` method on the `MessageProducer` (not the `Message` object!). The following code shows how this method is used by both the p2p and pub/sub message models:

```
// p2p setting the message priority to 9
QueueSender queueSender = QueueSession.createSender(someQueue);
queueSender.setPriority(9);

//pub/sub setting the message priority to 9
TopicPublisher topicPublisher = TopicSession.createPublisher(someTopic);
topicPublisher.setPriority(9);
```

Once a priority is established on a message producer (`QueueSender` or `TopicPublisher`), that priority will be used for all messages delivered from that producer, unless it is explicitly overridden. The priority of a specific message can be overridden during the send or publish operation. The following code shows how to override the priority of a message during the send and publish operations. In both cases, the priority is set to 3:

```
// p2p setting the priority on the send operation
QueueSender queueSender = QueueSession.createSender(someQueue);
queueSender.send(message,DeliveryMode.PERSISTENT, 3, 0);

// pub/sub setting the priority on the send operation
TopicPublisher topicPublisher = TopicSession.createPublisher(someTopic);
topicPublisher.publish(message,DeliveryMode.PERSISTENT, 3, 0);
```

There are two basic categories of message priorities: levels 0–4 are gradations of normal priority; levels 5–9 are gradations of expedited priority. Message servers are not required to enforce message ordering based on the `JMSPriority` header, but they should attempt to deliver expedited messages before normal messages.

The `JMSPriority` header is set automatically when the message is delivered. It can be read by JMS clients using the `Message.getJMSPriority()` method, but the accessor method is mostly used by message servers when routing messages.

JMSReplyTo

Purpose: Routing

In some cases, a message producer may want the consumers to reply to a message. The `JMSReplyTo` header indicates which destination, if any, a JMS consumer should reply to. The `JMSReplyTo` header is set explicitly by the JMS client; its contents will be a `javax.jms.Destination` object (either `Topic` or `Queue`).

In some cases, the JMS client will want the message consumers to reply to a temporary topic or queue set up by the JMS client. Here is an example of a pub/sub JMS client that creates a temporary topic and uses its `Topic` object identifier as a `JMSReplyTo` header:

```
TopicSession session =
connection.createTopicSession(false, Session.AUTO_ACKNOWLEDGE);
...
Topic tempTopic = session.createTemporaryTopic();
...

TextMessage message = session.createTextMessage();
message.setText(text);
message.setJMSReplyTo(tempTopic);
publisher.publish(message);
```

When a JMS message consumer receives a message that includes a `JMSReplyTo` destination, it can reply using that destination. A JMS consumer is not required to send a reply, but in some JMS applications, clients are programmed to do so. Here is an example of a JMS consumer that uses the `JMSReplyTo` header on a received message to send a reply. In this case, the subscriber will simple send an acknowledgment back to the publisher indicating it received the message:

```
Topic chatTopic = ... get topic from somewhere
...
// Publisher is created without a specified Topic
TopicPublisher publisher = session.createPublisher(null);
...

public void onMessage(Message message){
    try {
        TextMessage textMessage = (TextMessage)message;
        Topic replyTopic = (Topic)textMessage.getJMSReplyTo();
        TextMessage replyMessage = session.createTextMessage("Received Message...");
        publisher.publish(replyTopic, replyMessage);
    } catch (JMSException jmse){jmse.printStackTrace();}
}
```

The `JMSReplyTo` destination set by the message producer can be any destination in the messaging system. Using other established topics or queues allows the message producer to express routing preferences for the message itself or for replies to that message. Typically, this kind of routing is used in workflow applications. In a workflow application, a message represents some task that is processed one step at a time by several JMS clients—possibly over days. For example, an order message might be processed by sales first, then inventory, then shipping, and finally accounts receivable. When each JMS client (sales, inventory, shipping, or accounts receivable) is finished processing the order data, it could use the `JMSReplyTo` address to deliver the message to the next step.

JMSCorrelationID

Purpose: Routing

The `JMSCorrelationID` provides a header for associating the current message with some previous message or application-specific ID. In most cases, the `JMSCorrelationID` will be used to tag a message as a reply to a previous message. The following code shows how the

JMSCorrelationID is set and used along with the JMSReplyTo and JMSMessageID headers to send a reply to a message:

```
public void onMessage(Message message){
    try {
        TextMessage textMessage = (TextMessage)message;
        Queue replyQueue = (Queue)textMessage.getJMSReplyTo();

        Message replyMessage = session.createMessage();
        replyMessage.setJMSCorrelationID(message.getJMSMessageID());
        sender.send(replyQueue, replyMessage);
    } catch (JMSException jmse){jmse.printStackTrace();}
}
```

When the JMS client receives the reply message, it can match the JMSCorrelationID of the new message with the corresponding JMSMessageID of the message it sent, so that it knows which message received a reply. The JMSCorrelationID can be any value, not just a JMSMessageID. The JMSCorrelationID header is often used with application-specific identifiers. Our example in Chapter 4 uses the JMSCorrelationID as a way of identifying the sender. The important thing to remember, however, is that the JMSCorrelationID does not have to be a JMSMessageID, although it frequently is. If you decide to use your own ID, be aware that you should not start an application-specific JMSCorrelationID with ID:. That prefix is reserved for IDs generated by JMS providers.

The methods for accessing and mutating the JMSCorrelationID come in two forms: a String form and an AsBytes form. The String-based header is the most common and must be supported by JMS providers. The AsBytes method, which is based on a byte array, is an optional feature that JMS providers do not have to support. It's used for setting the JMSCorrelationID to some native JMS provider correlation ID:

```
Message message = queueSession.createMessage();
byte [] byteArray = ... set to some JMS specific byte array
...
message.setJMSCorrelationIDAsBytes(byteArray);
sender.send(message);
```

If the JMS provider supports messaging exchanges with a legacy messaging system that uses a native form of the correlation ID, the AsBytes method will be useful. If the AsBytes form is not supported, setJMSCorrelationIDAsBytes() throws a java.lang.UnsupportedOperationException.

JMSType

Purpose: Identification

JMSType is an optional header set by the JMS client. Its name is somewhat misleading because it has nothing to do with the type of message being sent (BytesMessage, MapMessage, etc.). Its main purpose is to identify the message structure and type of payload; it is only supported by a couple of vendors.

Some MOM systems (e.g., IBM's WebSphere MQ) treat the message body as uninterpreted bytes and provide applications with a simple way of labeling the body (the message type). So the message type header can be useful when exchanging messages with non-JMS clients that require this type of information to process the payload.

Other MOM systems (e.g., Sun's JMQ) directly tie each message to some form of external message schema, and the message type is the link. These MOM systems require the message type because they provide metadata services bound to it.

In addition, the `JMSType` might be used on a application level. For example, a messaging application that uses XML as its message payload might use the `JMSType` to keep track of which XML DTD the message payload conforms to. However, since the `JMSType` can possibly be used by JMS vendors, it would be much safer to use application properties, which are discussed in detail in Appendix C.

APPENDIX C

Message Properties

Message properties are additional headers that can be assigned to a message. They provide the application developer or JMS vendor with the ability to attach more information to a message. The Message interface provides several accessor and mutator methods for reading and writing properties. Properties can have a String value, or one of several primitive (boolean, byte, short, int, long, float, double) values. The naming of properties, together with their values and conversion rules, are strictly defined by JMS.

Property Names

Properties are name-value pairs. The name, called the identifier, can be just about any String that is a valid identifier in the Java language. With a couple of exceptions, the rules that apply to naming a property are the same as those that apply to the naming of variables. One difference between a JMS property name and a Java variable name is that a property name can be any length. In addition, property names are prohibited from using one of the message selector reserved words. These words include NOT, AND, OR, BETWEEN, LIKE, IN, IS, NULL, TRUE, and FALSE.

The property names used in JMS-defined properties and provider-specific properties use predefined prefixes. These prefixes (JMSX and JMS_) may not be used for application property names.

Property Values

Property values can be any boolean, byte, short, int, long, float, double, or String. The javax.jms.Message interface provides accessor and mutator methods for each of these property value types. Here is the portion of the Message interface definition that shows these methods:

```
package javax.jms;

public interface Message {
```

```
    public String getStringProperty(String name)
        throws JMSException, MessageFormatException;
    public void setStringProperty(String name, String value)
        throws JMSException, MessageNotWriteableException;

    public int getIntProperty(String name)
        throws JMSException, MessageFormatException;
    public void setIntProperty(String name, int value)
        throws JMSException, MessageNotWriteableException;

    public boolean getBooleanProperty(String name)
        throws JMSException, MessageFormatException;
    public void setBooleanProperty(String name, boolean value)
        throws JMSException, MessageNotWriteableException;

    public double getDoubleProperty(String name)
        throws JMSException, MessageFormatException;
    public void setDoubleProperty(String name, double value)
        throws JMSException, MessageNotWriteableException;

    public float getFloatProperty(String name)
        throws JMSException, MessageFormatException;
    public void setFloatProperty(String name, float value)
        throws JMSException, MessageNotWriteableException;

    public byte getByteProperty(String name)
        throws JMSException, MessageFormatException;
    public void setByteProperty(String name, byte value)
        throws JMSException, MessageNotWriteableException;

    public long getLongProperty(String name)
        throws JMSException, MessageFormatException;
    public void setLongPreperty(String name, long value)
        throws JMSException, MessageNotWriteableException;

    public short getShortProperty(String name)
        throws JMSException, MessageFormatException;
    public void setShortProperty(String name, short value)
        throws JMSException, MessageNotWriteableException;

    public Object getObjectProperty(String name)
        throws JMSException, MessageFormatException;
    public void setObjectProperty(String name, Object value)
        throws JMSException, MessageNotWriteableException;

    public void clearProperties()
        throws JMSException;
    public Enumeration getPropertyNames()
        throws JMSException;
    public boolean propertyExists(String name)
        throws JMSException;
    ...
}
```

The following code shows how a JMS client might produce and consume messages with properties that have primitive values:

```
// A message producer writes the properties
message.setStringProperty("Username","William");

message.setDoubleProperty("Limit", 33456.72);

message.setBooleanProperty("IsApproved",true);
publisher.publish(message);

...
// A message consumer reads the properties
String name = message.getStringProperty("Username");

double limit = message.getDoubleProperty("Limit");

boolean isApproved = message.getBooleanProperty("IsApproved");
```

The `Object` property methods that are defined in the `Message` interface (`setObjectProperty()` and `getObjectProperty()`) are also used for properties, but they don't give you as much functionality as their names suggest. Only the primitive wrappers that correspond to the allowed primitive types and the `String` type can be used by the `Object` property methods. Attempting to use any other `Object` type will result in a `javax.jms.MessageFormatException`. Here is an example of how the `Object` property methods are used to set and access properties in a message:

```
// A message producer writes the properties
String username = "William";
Double limit = new Double(33456.72);
Boolean isApproved = new Boolean(true);

message.setObjectProperty("Username",username);

message.setObjectProperty("Limit", limit);

message.setObjectProperty("IsApproved",isApproved);
publisher.publish(message);

...
// A message consumer reads the properties
String name = (String)message.getObjectProperty("username");

Double limit = (Double)message.getObjectProperty("Limit");

Boolean isApproved = (Boolean)message.getObjectProperty("IsApproved");
```

Immutable Properties

Once a message is sent, its properties become read-only; the properties cannot be changed. While consumers can read the properties using the property accessor methods (`get<TYPE>Property()`), they cannot modify the properties using any of the mutator methods (set<TYPE>Property()). If the consumer attempts to set a property, the mutator method throws a javax.jms.MessageNotWriteableException.

Once a message is received, the only way its properties can be changed is by clearing out all the properties using the clearProperties() method. This removes all the properties from the message so that new ones can be added. Individual properties cannot be modified or removed once a message is sent.

Property Value Conversion

The JMS specification defines rules for conversion of property values, so that, for example, a property value of type int can be read as a long:

```
// Set the property "Age" as an int value
message.setIntProperty("Age", 72);
...
// Read the property "Age" as a long is legal
long age = message.getLongProperty("Age");
```

The conversion rules are fairly simple, as shown in Table C-1. A property value can be set as one primitive type or String, and read as one of the other value types.

Table C-1. Property type conversions

Message.set<TYPE>Property()	Message.get<TYPE>Property()
boolean	boolean, String
byte	byte, short, int, long, String
short	short, int, long, String
int	int, long, String
long	long, String
float	float, double, String
double	double, String
String	String, boolean, byte, short, int, long, float, double

Each of the accessor methods (get<TYPE>Property()) can throw the MessageFormatException. The MessageFormatException is thrown by the accessor methods to indicate that the original type could not be converted to the type requested. The MessageFormatException might be thrown if, for example, a JMS client attempted to read a float property as an int.

String values can be converted to any primitive type, provided the String is formatted correctly:

```
Message message = topicSession.createMessage();

// Set the property "Weight" as a String value
message.setStringProperty("Weight","240.00");
```

```
// Set the property "IsProgrammer" as a String value
message.setStringProperty("IsProgrammer", "true");
...

// Read the property "Weight" as a float type
float weight = message.getFloatProperty("Weight");

// Read the property "IsProgrammer" as a boolean type
boolean isProgrammer = message.getBooleanProperty("IsProgrammer");
```

If the String value cannot be converted to the primitive type requested, a java.lang.NumberFormatException is thrown. Any property can be accessed as a String using the getStringProperty() method; all the primitive types can be converted to a String value.

The getObjectProperty() returns the appropriate object wrapper for that property. For example, an int can be retrieved by the message consumer as a java.lang.Integer object. Any property that is set using the setObjectProperty() method can also be accessed using the primitive property accessors; the conversion rules outlined in Table C-1 apply. The following code shows two properties (Age and Weight) that are set using primitive and Object property methods. The properties are later accessed using the Object, primitive, and String accessors:

```
Message message = topicSession.createMessage();

// Set the property "Weight" as a float value
message.setFloatProperty("Weight",240.00);

// Set the property "Age" as an Integer value
Integer age = new Integer(72);
message.setObjectProperty("Age", age);
...

// Read the property "Weight" as a java.lang.Float type
Float weight1 = (Float)message.getObjectProperty("Weight");

// Read the property "Weight" as a float type
float weight2 = message.getFloatProperty("Weight");

// Read the property "Age" as an Object type
Integer age1 = (Integer)message.getObjectProperty("Age");

// Read the property "Age" as a long is legal
long age2 = message.getLongProperty("Age");
```

Nonexistent Properties

If a JMS client attempts to access a nonexistent property using getObjectProperty(), null is returned. The rest of the property methods attempt to convert the null value to the requested type using the valueOf() operations. This results in some interesting behavior. The getStringProperty() returns a null or possibly an empty String ("")

depending on the implementation. The `getBooleanProperty()` method returns `false` for `null` values, while the other primitive property methods throw the `java.lang.NumberFormatException`.

The `propertyExists()` method can be used to avoid erroneous values or exceptions for properties that have not been set on the message. Here is an example of how it's used:

```
if (message.propertyExists("Age"))
   age = message.getIntProperty("Age");
}
```

Property Iteration

The `getPropertyNames()` method in the `Message` interface can be used to obtain an `Enumeration` of all the property names contained in the message. These names can then be used to obtain the property values using the property accessor methods. The following code shows how you might use this `Enumeration` to print all the property values:

```
public void onMessage(Message message) {
    Enumeration propertyNames = message.getPropertyNames();
    while(propertyNames.hasMoreElements()){
       String name = (String)propertyNames.nextElement();
       Object value = getObjectProperty(name);
       System.out.println("\nname+" = "+value);
    }
}
```

JMS-Defined Properties

JMS-defined properties have the same characteristics as application properties, except that most of them are set automatically by the JMS provider when the message is sent. JMS-defined properties are basically optional JMS headers; vendors can choose to support none, some, or all of them. There are nine JMS-defined properties, each of which starts with "JMSX" in the property name. Note that there is no corresponding `getJMSX<property>()` and `setJMSX<property>()` methods on the `Message` object; these properties are accessed in the same manner application properties are. For example, the following code checks to see if an application ID has been set by the JMS provider:

```
public void onMessage(Message message) {
   if (message.propertyExists("JMSXAppID")) {
      String appId = message.getStringProperty("JMSXAppID");
   }
   ...
}
```

Optional JMS-Defined Properties

Here are the optional JMS-defined properties and their descriptions:

`JMSXUserID`
: This property is a `String` that is set automatically by the JMS provider when the message is sent. Some JMS providers can assign a client a user ID, which is the value associated with this property.

`JMSXAppID`
: This property is a `String` that is set automatically by the JMS provider when the message is sent. Some JMS providers can assign an identifier to a specific JMS application, which is a set of consumers and subscribers that communicate using a set of destinations.

`JMSXProducerTXID` *and* `JMSXConsumerTXID`
: Messages can be produced and consumed within a transaction. Every transaction in a system has a unique identity that can be obtained from the producer or consumer using these properties. The `JMSXProducerTXID` is set by the JMS provider when the message is sent, and the `JMSXConsumerTXID` is set by the JMS provider when the message is received.

`JMSXRcvTimestamp`
: This property is a primitive `long` value that is set automatically by the JMS provider when the message is received. It represents the UTC time (see the section "UTC" on page 271) that the message was received by the consumer.

`JMSXDeliveryCount`
: This property is an `int` that is set automatically by the JMS provider when the message is received. If a message is not properly acknowledged by a consumer it may be redelivered. This property keeps a tally of the number of times the message server attempts to deliver the message to that particular consumer.

`JMSXState`
: This property is an `int` that is set automatically by the JMS provider. The property is for use by repositories and JMS provider tools and is not available to either the consumer or producer—as a developer, you will never have access to this property. The property provides a standard way for a JMS provider to annotate the state of a message. States can be one of the following: 1 (waiting), 2 (ready), 3 (expired), or 4 (retained). This property can be safely ignored by most JMS developers, but an explanation of its purpose is provided for completeness.

The JMS-defined properties that are assigned when the message is received (`JMSXConsumerTXID`, `JMSXRcvTimestamp`, and `JMSXDeliveryCount`) are not available to the message's producer, but only available to the message consumer.

Group JMS-Defined Properties

While the bulk of JMSX properties are optional, the group properties are not optional; they must be supported by the JMS provider. The group properties allow a JMS client to group messages together and assign each message in the group with a sequence ID. Here are the group properties:

`JMSXGroupID`
: This property is a `String` that is set by the JMS client before the message is sent. It is the identity of the group to which the message belongs.

`JMSXGroupSeq`
: This property is a primitive `int` type that is set by the JMS client before the message is sent. It is the sequence number of the message within a group of messages.

Provider-Specific Properties

Every JMS provider can define a set of proprietary properties of any type. These properties can be set by the client or the provider automatically. Provider-specific properties must start with the prefix "JMS_" followed by the property name (`JMS_<vendor-property-name>`). The purpose of the provider-specific properties is to support proprietary vendor features. You will need to refer to your JMS vendor documentation to find out what properties (if any) are supported by that particular vendor.

APPENDIX D

Installing and Configuring ActiveMQ

ActiveMQ is an enterprise-level open source messaging provider that supports JMS (as well as numerous native API's), making it a popular choice among messaging providers. It provides JMS clients with heterogeneous integration to a number of other platforms, including C++, C, C#, and Ruby (to name a few).

Although you can use any JMS provider that supports JNDI and JMS 1.1 to run the examples in this book, in this appendix we have provided the setup and configuration for using ActiveMQ. All of the code examples in the book can be run by using ActiveMQ "straight out of the box," with some minimal configuration changes. The following sections detail the configuration needed to run the examples using ActiveMQ version 5.2.0.

Installing ActiveMQ

The examples in this book use the basic out-of-the-box installation of ActiveMQ version 5.2.0. You can download ActiveMQ from the website at *http://activemq.apache.org*. Once downloaded, simply unzip or untar the compressed file.

Configuring ActiveMQ for JNDI

The examples in this book all use JNDI to connect to the JMS provider and obtain the JMS destinations (queues and topics). Therefore, you will need to create a *jndi.properties* file for each example. This property file contains the connection information, connection factory names, and destination names used by the sample code in each chapter. The code in this book is designed to be run with a centralized broker model, meaning that there is a separate and distinct JMS server running in its own JVM that JMS clients connect to.

In keeping with the spirit of the Java platform, we chose to use JNDI in the examples to make the source code JMS provider agnostic, allowing you to use any JMS provider supporting JNDI (as most do). In general, the *jndi.properties* file for the sample code in the book will require six properties to be set:

`java.naming.factory.initial`
: The initial context factory specific to each provider

`java.naming.provider.url`
: The protocol, address, and port of the JMS provider

`java.naming.security.principal`
: The user ID used to connect to the JMS provider

`java.naming.security.credentials`
: The password used to connect to the JMS provider

`connectionFactoryNames`
: The name(s) of the connection factory used to obtain JMS connections

`topic.<topicname>` (or `queue.<queuename>`)
: The JNDI name of the destination object (queue or topic)

ActiveMQ contains a configuration file located in the *ACTIVEMQ_HOME/conf* directory called *activemq.xml*. This file contains most of the configuration parameters needed for configuring and running ActiveMQ. Since the examples use the base configuration for ActiveMQ, only the sections of the *activemq.xml* file that need to be modified from the original base file will be shown.

The following sections illustrate the details needed in the *jndi.properties* file and the *activemq.xml* configuration file for running the examples in each chapter.

Configuration For Chat Examples

To run the `Chat` application found in Chapter 2, you will need to create a *jndi.properties* file (located in your classpath) with the `connectionFactoryNames` property set to `TopicCF` and the `topic.topic1` property set to `jms.topic1`:

```
java.naming.factory.initial = org.apache.activemq.jndi.ActiveMQInitialContextFactory
java.naming.provider.url = tcp://localhost:61616
java.naming.security.principal=system
java.naming.security.credentials=manager

connectionFactoryNames = TopicCF
topic.topic1 = jms.topic1
```

The *jndi.properties* file also contains the JNDI connection information for the JMS provider. You will need to set the initial context factory class, provider URL, username, and password needed to connect to the JMS server. To run the examples using ActiveMQ you would set the initial context factory to `org.apache.activemq.jndi.ActiveMQInitialContextFactory` and the provider URL to `tcp://localhost:61616` (the default protocol, host, and port for ActiveMQ (as shown earlier).

You will also need to define the topic used by the `Chat` application, called `topic1`. This is defined this in the *activemq.xml* file located in the *ACTIVEMQ_HOME/conf* directory. You will need to add a `destinations` element to the base configuration as shown here:

```
...
<destinations>
   <topic name="topic1" physicalName="jms.topic1" />
</destinations>
...
```

In addition to the configuration changes indicated here, you will need to include the *activemq-all-5.2.0.jar* file in your classpath.

Configuration for P2P Examples

The configuration for running the code examples in Chapter 4 is similar to that for the `Chat` application previously. To run the `QBorrower` and `QLender` code, you will need to define two new queues, `LoanRequestQ` and `LoanResponseQ`. In addition to the connection properties (which are the same as the `Chat` application), the *jndi.properties* file should contain three additional properties, the `connectionFactoryNames` property (which is set to `QueueCF`), the `queue.LoanRequestQ` property (which is set to `jms.LoanRequestQ`), and finally the `queue.LoanResponseQ` property (which is set to `jms.LoanResponseQ`). The `jms` part of the name for the `jms.LoanRequestQ` and `jms.LoanResponseQ` destination property values is completely arbitrary; you could easily name them `mydestination.LoanRequestQ` and `mydestination.LoanResponseQ`. Whatever you choose for the destination names, just make sure the names in the *jndi.properties* file match the physical names specified in the *activemq.xml* file.

The *jndi.properties* file for the examples in Chapter 4 is as follows:

```
java.naming.factory.initial = org.apache.activemq.jndi.ActiveMQInitialContextFactory
java.naming.provider.url = tcp://localhost:61616
java.naming.security.principal=system
java.naming.security.credentials=manager

connectionFactoryNames = QueueCF
queue.LoanRequestQ = jms.LoanRequestQ
queue.LoanREsponseQ = jms.LoanResponseQ
```

You will also need to define the queues used by the `QBorrower` and `QLender` classes. These are defined in the *activemq.xml* file located in the *ACTIVEMQ_HOME/conf* directory. You will need to add a `destinations` element to the base configuration as shown here:

```
..
<destinations>
   <queue name="LoanRequestQ" physicalName="jms.LoanRequestQ" />
   <queue name="LoanResponseQ" physicalName="jms.LoanResponseQ" />
</destinations>
...
```

In addition to the configuration changes indicated here, you will need to include the *activemq-all-5.2.0.jar* file in your classpath.

Configuration for Pub/Sub Examples

The ActiveMQ configuration for running the `TBorrower` and the `TLender` classes is similar to the configuration specified in the `Chat` application, only the topic name is set to `RateTopic` rather than `topic1`. Everything else is the same.

The *jndi.properties* file for the examples in Chapter 5 is as follows:

```
java.naming.factory.initial = org.apache.activemq.jndi.ActiveMQInitialContextFactory
java.naming.provider.url = tcp://localhost:61616
java.naming.security.principal=system
java.naming.security.credentials=manager

connectionFactoryNames = TopicCF
topic.RateTopic = jms.RateTopic
```

You will also need to define the topic used by the `TBorrower` and `TLender` classes. This is defined in the *activemq.xml* file located in the *ACTIVEMQ_HOME/conf* directory. You will need to add a `destinations` element to the base configuration as shown here:

```
..
<destinations>
   <topic name="RateTopic" physicalName="jms.RateTopic" />
</destinations>
...
```

In addition to the configuration changes indicated here, you will need to include the *activemq-all-5.2.0.jar* file in your classpath.

Configuration for Spring JMS Examples

This section contains the ActiveMQ configuration required to run the Spring JMS examples in Chapter 9. Like the other examples found in this book, you will need to create a *jndi.properties* file (located in your classpath) with the `connectionFactoryNames` property set to `QueueCF` and the `queue.queue1` property set to `jms.queue1`:

```
java.naming.factory.initial = org.apache.activemq.jndi.ActiveMQInitialContextFactory
java.naming.provider.url = tcp://localhost:61616
java.naming.security.principal=system
java.naming.security.credentials=manager

connectionFactoryNames = QueueCF
queue.queue1 = jms.queue1
```

The *jndi.properties* file also contains the JNDI connection information for the JMS provider. You will need to set the initial context factory class, provider URL, username, and password needed to connect to the JMS server. To run the examples using ActiveMQ, you would set the initial context factory to `org.apache.activemq.jndi.ActiveMQInitialContextFactory` and the provider URL to `tcp://localhost:61616` (the default protocol, host, and port for ActiveMQ (as shown previously).

You will also need to define the queue used by the Spring JMS examples, called `queue1`. This is defined in the *activemq.xml* file by adding a `destinations` element as shown here:

```
..
<destinations>
   <queue name="queue1" physicalName="jms.queue1" />
</destinations>
...
```

In addition to the configuration changes indicated here, you will need to include the *activemq-all-5.2.0.jar* file in your classpath.

Index

Symbols and Numbers

A

We'd like to hear your suggestions for improving our indexes. Send email to *index@oreilly.com*.

B

C

D

E

F

G

H

I

J

K

L

M

N

O

P

Q

R

S

T

U

V

W

X

About the Authors

Mark Richards is an accomplished author and conference speaker working as a hands-on SOA and enterprise architect in the financial services industry. In addition to numerous published articles, he is the author of *Java Transaction Design Strategies* (C4Media), contributing author of *97 Things Every Software Architect Should Know* (O'Reilly), and contributing author of *No Fluff, Just Stuff Anthology* Volumes 1 and 2 (Pragmatic Bookshelf). He is a recognized authority on messaging, Service-Oriented Architecture, and transaction management. Mark is a regular speaker on the NFJS Software Symposium series and speaks at conferences around the world.

Richard Monson-Haefel is the author of the first five editions of *Enterprise Java Beans* (O'Reilly), the first edition of *Java Message Service* (O'Reilly), and is one of the world's leading experts and book authors on enterprise computing.

David A. Chappell is vice president and chief technologist for SOA at Oracle Corporation. He is well noted for authoring *Java Web Services* (O'Reilly), *Professional ebXML Foundations* (Wrox), and the first edition of *Java Message Service* (O'Reilly).

Colophon

The animal on the cover of *Java Message Service*, Second Edition, is a passenger pigeon (*Ectopistes migratorius*), an extinct species. In the mid-1800s, passenger pigeons were the most numerous birds in North America. Several flocks, each numbering more than two billion birds, lived in various habitats east of the Rocky Mountains. Flocks migrated en masse in search of food, without regard to season, and a good food source could keep a flock in one place for years at a time. John James Audubon observed that nearly the entire passenger pigeon population once stayed in Kentucky for several years and was seen nowhere else during this time.

Whole flocks roosted together in small areas, and the weight of so many birds—often up to 90 nests in a single tree—resulted in the destruction of forests, as tree limbs and even entire trees toppled. (The accumulated inches of bird dung on the ground didn't help.) Such roosting habits, combined with high infant mortality and the fact that female passenger pigeons laid a single egg in a flimsy nest, did not bode well for the long-term survival of the species.

It was humans harvesting passenger pigeons for food, however, that drove them to extinction. In 1855, a single operation was processing 18,000 birds per day! Not even Audubon himself was concerned that the pace might have an adverse effect on the birds' population, but the last passenger pigeon died in the Cincinnati Zoo in 1914.

The cover image is a 19th-century engraving from the Dover Pictorial Archive. The cover font is Adobe ITC Garamond. The text font is Linotype Birka; the heading font is Adobe Myriad Condensed; and the code font is LucasFont's TheSansMonoCondensed.

Related Titles from O'Reilly

Java

Ajax on Java

Ant: The Definitive Guide, *2nd Edition*

Better, Faster, Lighter Java

Beyond Java

Eclipse

Eclipse Cookbook

Eclipse IDE Pocket Guide

Enterprise JavaBeans 3.0, *5th Edition*

Hardcore Java

Head First Design Patterns

Head First Design Patterns Poster

Head First Java, *2nd Edition*

Head First Servlets & JSP

Head First EJB

Hibernate: A Developer's Notebook

J2EE Design Patterns

Java 5.0 Tiger: A Developer's Notebook

Java & XML Data Binding

Java & XML, *3rd Edition*

Java Cookbook, *2nd Edition*

Java Data Objects

Java Database Best Practices

Java Enterprise Best Practices

Java Enterprise in a Nutshell, *3rd Edition*

Java Examples in a Nutshell, *3rd Edition*

Java Extreme Programming Cookbook

Java Generics and Collections

Java in a Nutshell, *5th Edition*

Java I/O, *2nd Edition*

Java Management Extensions

Java Message Service

Java Network Programming, *3rd Edition*

Java NIO

Java Performance Tuning, *2nd Edition*

Java RMI

Java Security, *2nd Edition*

JavaServer Faces

JavaServer Pages, *3rd Edition*

Java Servlet & JSP Cookbook

Java Servlet Programming, *2nd Edition*

Java Swing, *2nd Edition*

Java Web Services in a Nutshell

JBoss: A Developer's Notebook

JBoss at Work: A Practical Guide

Learning Java, *3rd Edition*

Mac OS X for Java Geeks

Maven: A Developer's Notebook

Programming Jakarta Struts, *2nd Edition*

QuickTime for Java: A Developer's Notebook

Spring: A Developer's Notebook

Swing Hacks

Tomcat: The Definitive Guide, *2nd Edition*

WebLogic: The Definitive Guide

O'REILLY®